HARTLEPOOL BOROUGH
WITHDRAWN
LIBRARIES

D0321718

Thatching and Thatched Buildings

By the same author

Thatched Buildings of Dorset

Thatching and Thatched Buildings

Michael Billett

ROBERT HALE · LONDON

First published in Great Britain 1979
Reprinted 1982
Enlarged second edition 1988

Robert Hale Limited
Clerkenwell House
Clerkenwell Green
London EC1R 0HT

British Library Cataloguing in Publication Data

Billett, Michael
Thatching and thatched buildings.
— 2nd enl. ed.
1. England. Thatching 2. England
Thatched buildings
I. Title
695

ISBN 0-7090-3523-3

Printed in Great Britain by
St Edmundsbury Press, Bury St Edmunds, Suffolk
Bound by Woolnough Bookbinding Ltd

Contents

Illustrations

LIST OF PHOTOGRAPHS

PICTURE CREDITS

Highlands and Islands Development Board: 1. Wales Tourist Board Photo Library: 2. Roy J. Westlake: 3, 4, 10, 19, 21, 31. Richard Jemmett: 5, 6, 8, 12, 13, 20, 25, 26, 29. Geoffrey N. Wright: 7, 9, 11, 14, 16, 22, 24, 28, 30, 32. Joy Wotton: 15. John Baguley: 17. Eileen Preston: 18, 23, 27.

LIST OF LINE DRAWINGS

(1)
Historical Development of Thatch and Thatched Buildings

The use of thatch as a roof-covering material has very ancient origins. This was because nature had made available suitable plants which grew in most parts of the world and could be adapted into some form of roof covering. In the broad sense, the term 'thatch' implies any roof made of matter of vegetable origin. Coconut leaves are still utilized in many under-developed tropical areas of the world for the roofs of simple hut shelters. The palmetto or thatch palm of the West Indies yields large leaves which are frequently used for thatching. The wood of the tree is also sometimes employed as a thatch material. In England, thatch in various forms has been in use throughout many centuries. The precise material used has largely depended upon its ready local availability. The material may have been brushwood, turf, heather, broom, water reeds or straw derived from corn.

The art of thatching is probably the most ancient building craft still practised in England. It started when man first left his cave-dwelling existence and learned to grow crops and tend animals. It then became essential for him to build some form of shelter nearby for his own protection. During the Stone Age, shelters were covered with a crude form of thatch. The first simple outdoor shelter was probably constructed by piling cut branches around a standing pole, such as a tree. The circular-shaped structure was then covered with a rough thatch made of brushwood, bracken or turf. The practice continued through the Iron Age, still using a crude form of thatch over wooden supports. It is likely that in areas where timber may have been scarce, rough stone walls may also have been made to support the roof. A simple development in the roof construction was the cutting of a hole in the thatch, to allow the smoke from a fire lit within the hut to escape. Holes in the roof were therefore an early form of smoke flue. The entrances to the huts were made as small as possible, to prevent or delay attacks from intruders. Window spaces were not constructed.

In later times, the Romans occasionally used thatched roofs on their dwellings in country areas but, because of the fire danger, they came to prefer stone slabs and eventually tiles, which they manufactured. With the departure of the Romans from Britain, the early Saxon settlers discontinued the use of stone and built houses in perishable materials such as thatch, timber and wattle and daub. Wattle was a type of hurdling made from wood, probably hazel, which was traditionally preferred later. The daub was a coating of mud or clay smeared over the wattle. The Saxons left the Roman towns to decay and built new villages in the countryside. The villages were usually sited by streams and built on fertile soil, as was dictated by their good early theory of agriculture. The need for water usually caused the villages to be built in valleys, or mainly on the lower southern-facing slopes, where spring

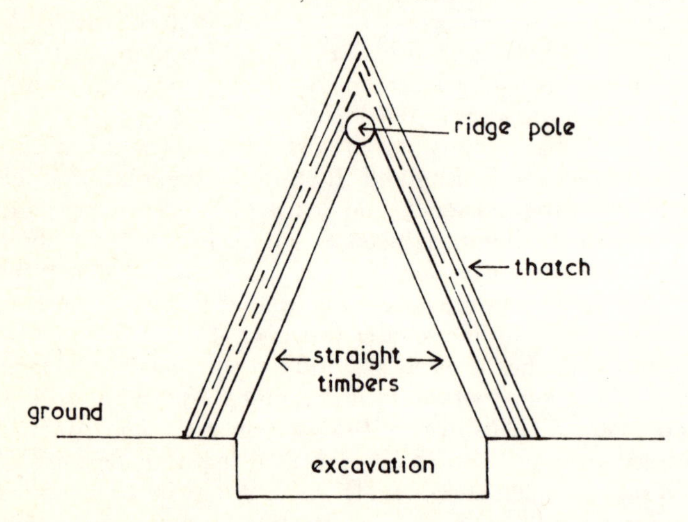

Fig.1 Saxon house (cross-section)

water was likely to be available. This type of village location in England is still in evidence, and it is in contrast to that found in other countries of Europe, which frequently have villages sited on the tops of hills. This is only occasionally found in England, where a good vantage-point was originally required for a particular observation or defence purpose.

The houses built by the early Saxons were normally made in such a fashion that they appeared to be all roof with no walls (Fig.1). In the early designs, a shallow excavation was dug approximating to the area of the proposed house. Two straight heavy timbers were then bound together and the ends set in two of the corners of the rectangular-shaped excavated area. Another similar set of timbers was placed at the two other end corners of the excavation. A strong timber pole was then lashed across to connect the then triangular-shaped end-pieces of the house. This connecting pole formed the ridge timber to support the roof, which was made of rough thatch, possibly rushes, laid on a support of brushwood or wattle and daub.

A hole was again cut in the roof to act as a smoke hole. The fuel used for the fires lit inside the Saxon houses was either wood or turfs. The atmosphere within the houses must have been smoky and foul. However, the fumes from the fires would have been appreciably more toxic if coal had been burned, instead of wood or turfs. Fortunately, coal was not in use as a fuel in these early days, before wattle smoke-shafts had been evolved and eventually chimneys at a much later date in history. The all-roof-with-no-wall design of the early Saxon houses considerably restricted the head space within the house. Walls of timber were later constructed to raise the roof level. Timber frames were also utilized and filled with wattle and daub, wood or stones. Eventually this led to the use of curved timbers for the house end-pieces and the evolution of the early type of cruck roof construction. The cruck gave rise to an arch shape due to the lashing together of the curved timbers (Fig.2). The very early cruck houses were thatched, and the design gave additional head room. The thatch material used would have been rushes, turf, ling or straw, depending on the area.

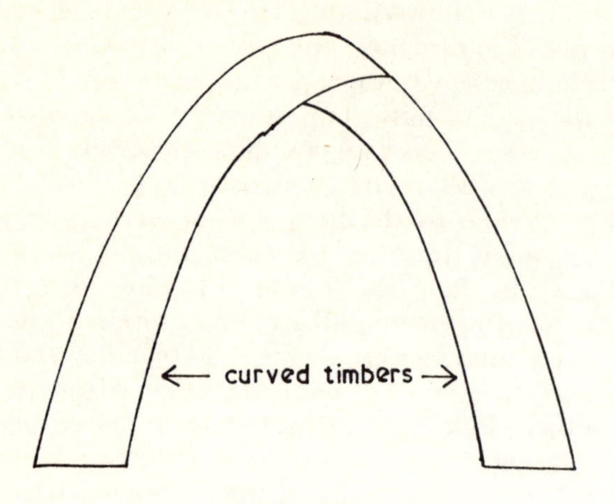

Fig.2 Cruck

The Saxon houses were of a barnlike nature, and families shared them with their domestic animals and cattle, which were kept in separate bays. The early building of the houses in shallow excavations must have led to considerable dampness and flooding. Eventually, this building practice discontinued – later houses were built on slightly sloping land and the animals were kept at the lower end of the house slope for obvious drainage reasons. Local available material was used for the thatching. This was often water reeds, and the art of thatching with this material, as we know it today, probably dates back to at least the tenth century. Therefore, present-day methods of reed thatching have not drastically altered since the later Saxon era. During the following Norman period, stone buildings were constructed, and many examples are still standing today, due to the more robust nature of some of them. Although the majority of the population lived in the countryside, the building of houses in small townships was also taking place. However, most people under the feudal system still lived in humble dwellings constructed of perishable materials, such as thatch and wattle and daub.

The fire risk with thatch, in early times, was always particularly great because open fires were still lit inside the dwelling, and the only exit for the smoke was a hole cut in the thatched roof. The fire problem assumed even greater seriousness when thatch and wood were employed in the early Middle Ages as building materials for houses in close proximity to one another in the towns. In the year 1212, the use of thatch for the roof construction of new buildings was banned in London. Stone tiles and sometimes lead were used as alternative roofing materials. It was also decreed that existing thatched roofs should be protectively coated with a layer of a crude lime-based plaster to reduce the risk of ignition. This practice of whitewashing thatched roofs to increase their fire resistance spread to other parts of the country. Even today, it is still possible to see the occasional whitewashed thatched roof in parts of Yorkshire and even sometimes in Wales and Scotland.

Special tools and hooks were made so that burning thatch material could be quickly pulled away from the roof. The fire-hooks were attached to long poles, and the implements were heavy and cumbersome to handle. Later it became the practice to install iron rings on the eaves of the houses, which allowed ropes to be inserted through the rings to assist the pulling up and the supporting of the weight of the hook and pole. The incidence of many fires in the Middle Ages led to the gradual development of chimneys in England during the ensuing centuries. The first tall chimney-stacks were introduced at the end of the twelfth century, but they were not generally used until much later. In country areas, fires were not such a problem as in the towns because the houses were more isolated. Lightning sometimes caused thatch fires, but the villagers would normally be able to cope, before the complete destruction of the house. Oak was frequently used, especially in later times, for the rafter construction, and due to this wood's resistance to fire, it would often be merely charred after the burning thatch had been pulled away by a long hook. Well or river water was the only other available means of fire fighting, as mains tap water was, of course, not available in these times. After a fire had been extinguished, the roof was thatched again by the villagers. The furniture and other items inside the buildings were relatively simple: they were manufactured from products which were not as combustible as those used in modern-day furnishing; wallpaper, curtains, synthetic paint and varnish were also absent. The general fire risk was therefore much less.

On the other hand, much wood was used in the house construction, in addition to the wattle support for the under thatch on the oak rafters. Timber was much utilized for wall building in the twelfth and thirteenth centuries, even in areas such as the Cotswolds where large quantities of limestone were available. This was because wooden houses were easier and cheaper to build. The combination of wood and thatch therefore always presented a potential fire risk. (It was not until the end of the thirteenth century that local stone was used widely to replace timber in the Cotswolds.)

In the following centuries, and even after limited quantities of fired bricks became available in certain areas in the seventeenth century, local materials continued to be used for the walls of thatched buildings. Country areas could not easily obtain building materials from outside regions, and construction was therefore carried on using old-established methods. Cob, a mixture of clay, straw and gravel, or sometimes mud, straw and horsehair, was much used for wall construction, particularly in Devon and also in Cornwall and Somerset. A very large number of these cob buildings, white or colour-washed, can be seen today. It was found that cob was prone to deteriorate if exposed to damp, but if adequately protected it would last for centuries. Cob walls were therefore constructed on the top of a stone foundation built at ground level to lessen the effect of rising ground dampness. An overhanging thatched roof adequately protected the cob walls from falling rainwater.

Stone ashlar and chalk ashlar, materials cut into the shape of large building blocks, were much used in Dorset and Somerset for wall construction. Chalk ashlar also lasted for centuries if adequately protected from water attack. This was found to be particularly so if the walls were covered with a thatched roof to keep them perfectly dry and prevent water gaining access into the top of the walls. Local stones were often used where they were available in various parts of

England. However, if timber was more abundant, then this was utilized. Combinations of timber and local stone were also employed. The thatch materials used for the roof construction were again those most readily available. These were usually either marsh water reeds, ling, straw or sometimes turfs. Thatch was therefore predominantly used in the country areas where these materials were abundant. If stone was more readily to hand, then stone tiles would be utilized instead of thatch.

In the medieval era, a newer, better type of thatched house was being built for the yeomen farmers who were owners of land. They, and the various craftsmen of the time, wanted houses that were better and larger than those inhabited by the peasants. However, the yeomen houses were not so grand as the manor houses. During this time in history, thatch was also used as a temporary roof covering for churches, especially those constructed in East Anglia and, to a lesser extent, in the West Country. There are still approximately thirty thatched churches surviving today.

In early Tudor times, at the beginning of the sixteenth century, farmers built their own houses in the village and not on the site of their farm buildings. It was only later that thatched farmhouses could be built in the centre of the farm, when the enclosing of land became permissible. There are many of these thatched farmers' and also yeomen houses still surviving and being lived in today. They are normally much larger than those originally constructed for the farm workers. During this era, it was also popular to build the type known as the medieval long house. As the name implies, these houses were extremely long in comparison with their width. It was the practice to support the thatched roof over many huge timbered beams and to build large fireplaces. At this time, thatch was not so frequently used for manor house roof construction because the wealthy were coming to think of it as an inferior material.

In Elizabethan times, during the second half of the sixteenth century, the population of England had become approximately four million, of whom three quarters lived in villages rather than towns. However, the rough pattern for the typical English village had been set before this, in medieval times. Particularly in southern England, there was the church, the manor house, farms, thatched dwellings and often a village green and pond. Many villages were still sited on the original Saxon locations. It was during the Elizabethan period that the village really developed into a relatively self-sufficient unit. There was an impetus to expand food production to feed the increasing population of England. Many additional simple thatched cottages were therefore built in the countryside, especially where corn was being produced, and this also yielded straw for thatching. A few more elaborately thatched houses were also being built during the Elizabethan period. It was at this time that the timber-framed house, with various in-fillings, became popular. This was especially so in Norfolk and Suffolk but also in parts of Warwickshire and Worcestershire.

The houses built in the sixteenth century were predominantly square or oblong in shape. When they were thatched, it was the custom to lay the thatch over all the four walls in a hipped fashion, with eaves all around the house (Fig.3). The thatch was swept around the four corners of the house, as it was difficult to construct a satisfactory gable end

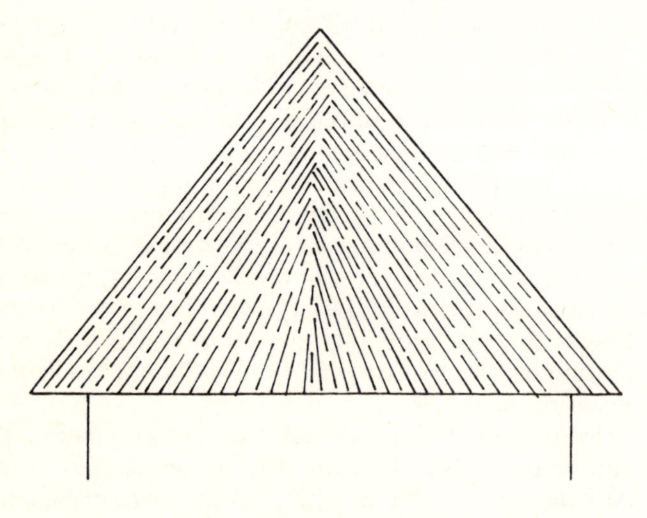

Fig.3 Hipped roof end

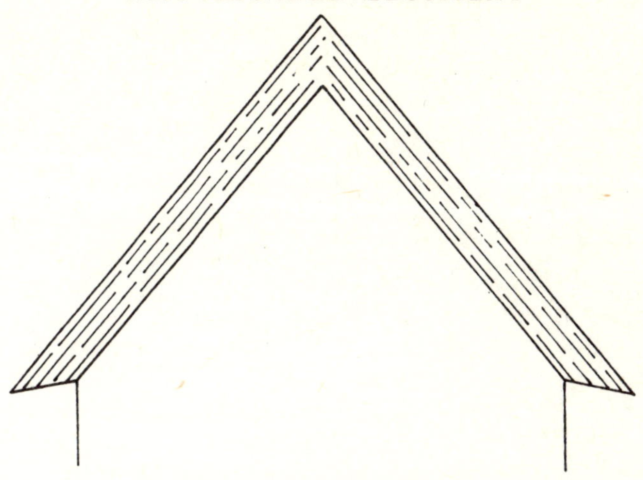

Fig.4 Gabled roof end

in thatch (Fig.4). The thatch under an exposed gable end was thought prone to be lifted by the wind, with possible deterioration of the roof. In England, thatched roofs had always been made steep, mainly to shed snow so that its accumulated weight would not damage the thatch. However, head room in the houses was restricted, particularly near the walls, due to the inclination of the roof (Fig.5). Most thatched houses in the country areas were of single-storey construction with no entrance to a loft, which accentuated the space problem. In order to increase the head room, it became the practice to build houses in the medieval long house style with the house ends particularly steeply thatched (Fig.6). In the second half of the sixteenth century, the people who were able to afford something better, had a top storey built to obtain additional space. These Tudor-type buildings were relatively small, and the upper storey was given an overhang. The roofs were frequently steeply thatched and hipped, although by this time satisfactory gable ends could be made.

A further variation later appeared in the south-east of England with the development of the Sussex hip (Fig.7). This allowed the thatch to be swept upwards around an attic window space in an end wall. Many examples of thatched

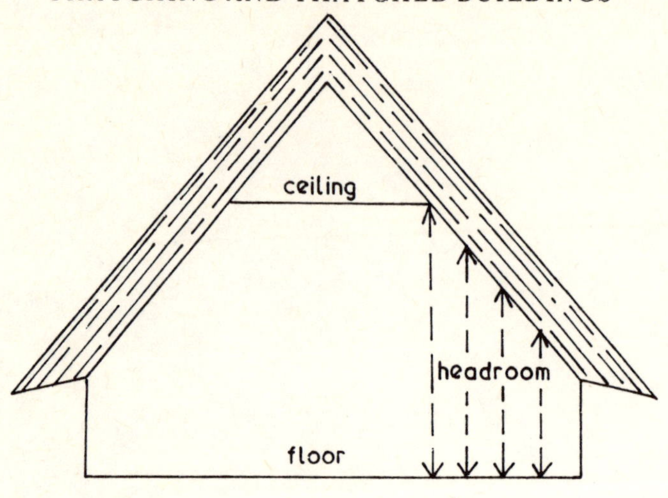

Fig.5 Roof inclination

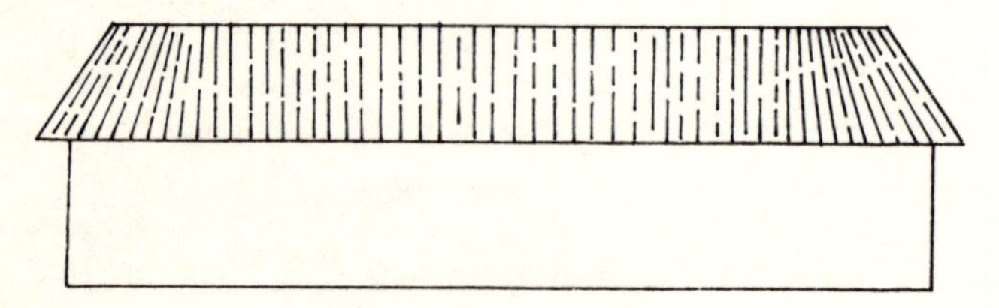

Fig.6 Long House

houses of this type can still be seen. It was during the seventeenth century that the widespread use of glass windows became popular. Before this time, any window spaces in houses had been kept as small as possible to retain warmth. In 1695 a window tax was introduced in England, due to the increasing areas of window glass being used. The tax had the effect of reducing window spaces again, as people blocked up some of their windows to avoid the tax. It was also during the seventeenth century that the dormer window was introduced into house designs. With thatched buildings, a wall section was often raised to allow the construction of a dormer, so that the window could be placed vertically in the sloping roof. The thatch was then gently swept around the top of the window, thereby assuming the shape of an eyebrow (Fig.8).

From the seventeenth to the early part of the nineteenth century, the use of manufactured tiles for roofs became popular in England. Thatch was then predominantly found only on the more humble dwellings which housed the poor. Thatch was utilized for these because it was cheap and also

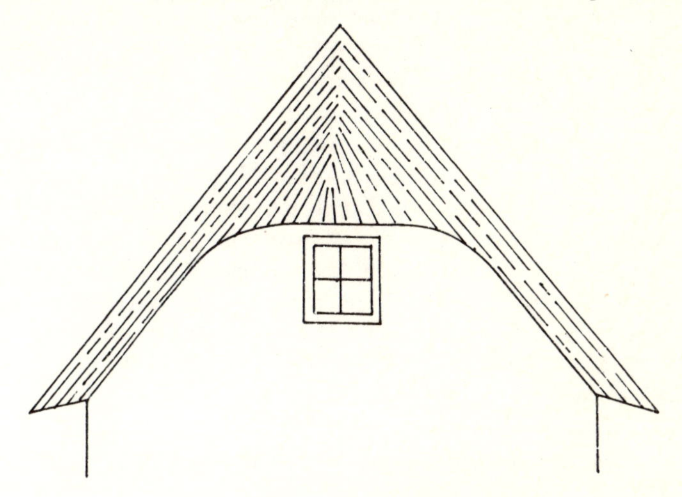

Fig.7 Sussex hip

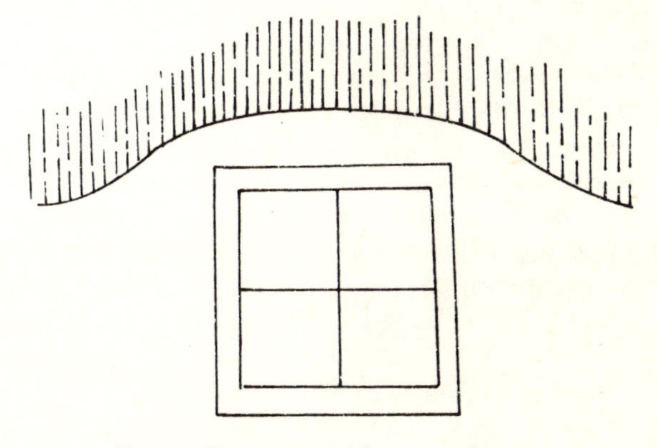

Fig.8 Dormer window – eyebrow

because it was relatively light in weight. It could be used as a roofing material over dwellings which had walls constructed of cheap locally manufactured materials, such as cob. The walls were built directly on a stone layer on the ground without dug foundations. It was not fashionable in these times to live under a thatched roof, and people went to great lengths of pretence, such as having the front roof tiled, where it could be viewed from the road, and the back one thatched. The situation changed in the first half of the nineteenth century when slates were introduced from Wales, as a cheap commercially available product. The quarrying of slates in large quantities eventually became just as important an industry to North Wales as coal mining to South Wales. With the growth of the railways, it was possible to distribute them widely throughout the whole United Kingdom. In comparison, thatch became a more expensive product and a fashionable material again with most classes of society. Wheat straw, in particular, had become relatively expensive due to shortages during the earlier Napoleonic Wars.

In the early part of the nineteenth century, thatch had again become fashionable enough for the wealthy to build whole country estates containing thatched buildings. Many planned, integrated villages were also built consisting of nearly identical thatched cottages. Thatched dwellings in the *cottage ornée* style became popular. This design denoted a small building deliberately built to capture the careless rustic style, as thought most appropriate to England. Large country houses were also designed in a cottage style. In such designs, a steeply inclined thatched roof structure was supported by vertical heavy timbers, which were ornamentally set around the outside of the house in front of the main wall structure. This type of thatched building with rustic pillars was exceptionally popular in the Isle of Wight. In the early part of the nineteenth century, it also became fashionable to build rather ornate arch-shaped doorways in some thatched farmhouses. However, the practice did not retain its popularity for long because the architectural style clashed with the aesthetic appeal of the thatched roof.

Preferences for particular shapes and styles of roof in

various areas gradually became more pronounced. In Norfolk, and also in East Anglia generally, the roofs were built particularly highly pitched with steep, sharp-edged gables. In the West Country, a more rounded chunky appearance had evolved. In Essex, it became a fashion to shape the gable ends with a rounded or barge effect. In the Cotswolds, the development was normally towards gable-ended rather than hipped roofs. This was in contrast to Kent and also to parts of Buckinghamshire and Wessex, where the hipped shape became particularly popular. Preferred roof shapes were not just confined to specific regions of England. For example, in the south and south-east of Ireland there still remains a predominance of thatched hipped-roof houses. In contrast to this, the northern and western parts of the country possess mainly the gable type of roof. As in England, old farmhouses possessing thatched roofs can still be seen throughout Ireland. They are normally single-storey long buildings with whitewashed stone walls, although they are sometimes colour-washed. Occasionally, two-storey thatched houses may be seen, but these are rare.

In country areas, thatch was used for many centuries for the protection of buildings other than inhabited dwellings. For example, thatched barns were commonly built, and some of these possessed huge thatched roofs, especially those constructed in the south-east of England. The use of thatch on barns sometimes presented problems: the thatched roof was able to cope adequately with wind pressure on the outside of the roof, but if a barn door was left open and the wind pressure hit the inside of the roof, then parts of the thatch were likely to be lifted and carried away. Thatched barns are still encountered on some farms and are a magnificent distinguishing feature and reminder of the old method of storing grain.

The large sizes and roof areas of barns now make them a very expensive proposition to maintain and re-thatch. Many farmers may face expenditure of several thousands of pounds to preserve even medium-sized barns. It would be a pity if such buildings were allowed to disappear from the English countryside because of lack of finance. Fortunately, many are now 'listed' or 'scheduled buildings', and therefore it is sometimes possible for financial assistance towards maintenance to be obtained with a government or local authority grant. There are also many old thatched tithe barns scattered throughout England. These were originally used to store the percentage contributions of grain which each parish donated to raise money for paying the tithes to maintain the church. These large old tithe barns usually had two doors at the opposite ends of the building. Wagons could then readily enter a barn to unload the grain and then leave by the other exit, thus not having to turn in the middle of the barn. In Dorset, thatched roof corn-stores were often built close to farmers' barns. These stores had timber walls and were raised off the ground by placing them on a number of stone legs (Fig.9).

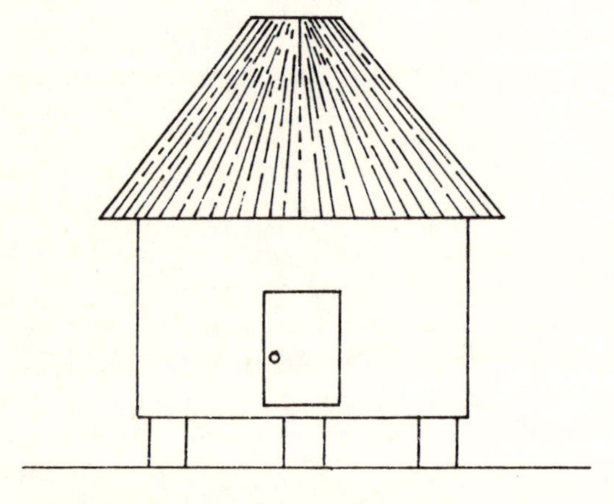

Fig.9 Corn-store

It was also common practice in England to thatch ricks and hay stacks. It was not a practicable proposition for the local thatcher to thatch a very large number of ricks on his own, as all the farmers required his services at the same time, at the end of the harvest season. This thatching operation was usually carried out by the farm workers, as well as by the thatcher, because the quality of the thatch required was not important, as it did not need to be as durable as a house roof thatch. Speed was also an essential factor in rick thatching, in order to beat the weather. The thatch on the ricks was normally secured in position with the use of plaited straw ropes and twisted hazel spars.

The top of the thatched rick or haystack was often ornamented with a corn dolly, which was usually made in the shape of a bird or animal. Corn dollies were used for many centuries as good-luck charms for the future harvest. They were traditionally made from the last sheaf of corn of the final wagon-load of the gathered summer harvest. They were kept throughout the winter, and the grain obtained from them was sown the following spring to ensure a bountiful harvest during the coming months. The idea was later modified so that the corn dolly was placed on the rick or hay stack roof. These finished ricks and hay stacks were very pleasing to the eye. However, for many years now, due to the advent of the combine harvester, they have been replaced by the Dutch barns with arched metal roofs. On scientific grounds, a modern Dutch barn, consisting of a metal roof on sturdy pillars, gives not only a rain-proof cover but also perfect ventilation for straw and hay. Nevertheless, the design lacks the aesthetic appeal of an old thatched barn, rick or hay stack.

Thatchers and farm workers often made straw ropes, for use in securing thatch, from a standing straw rick. This was done by using an instrument called a whimmer (Fig.10). The whimmer consisted basically of a metal hook, a holding handle and a turning handle. The hook would be pushed into

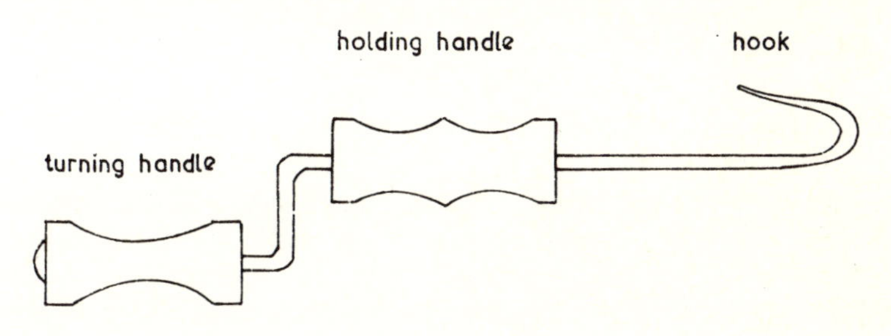

Fig.10 Whimmer

the straw rick, so that when it was withdrawn it would bring with it a small amount of straw. The person holding the whimmer would then walk slowly backwards (Fig.11). At the same time, one hand would clasp the holding handle and the other hand would turn the end handle in the manner of use of a carpenter's brace. The twisting and pulling action had the effect of fabricating and drawing out a straw rope from the rick. The rope could be made to any desired length, depending on the distance the operator walked backwards from the rick.

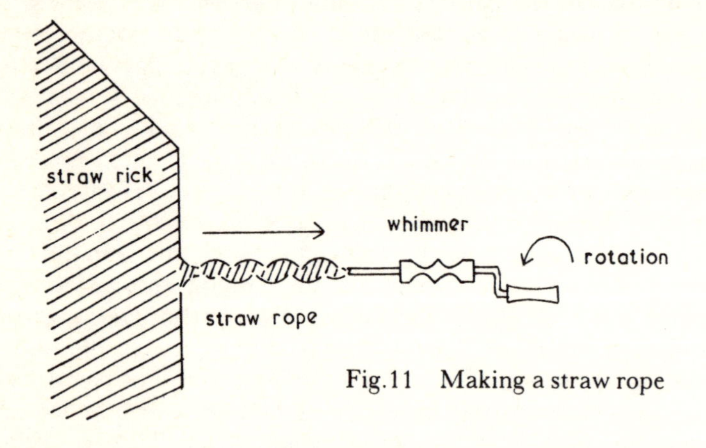

Fig.11 Making a straw rope

Thatch was once employed in the countryside to protect milk-stands and platforms from the rain. These used to be a familiar sight along many country roadways. Little thatched roofs or canopies were also often built over village notice-boards to protect the paper and prevent the ink from running in the rain. Another use for thatch in the past was the protection of walls, used as boundary markers, from disintegrating in the rain. Tops of walls were also capped with a thatch layer as a temporary measure against frost damage during their construction. The unusual sight of the thatched wall can still be seen in Wiltshire and also parts of Berkshire and Hampshire. Originally, these walls were made by pouring a mixture of water, chalk, flint and other ingredients into two previously erected pieces of wooden formwork, arranged vertically in parallel. These forms acted as the mould to maintain the wall shape, until the water had evaporated from the mixture, after which the formwork was removed, leaving the wall. The materials used to construct walls were not particularly stable under adverse weather conditions. It was therefore found necessary to protect the walls from the rain by the construction of a roof layer of thatch over their entire lengths. If the walls had not been protected in this way, they would have disintegrated long ago and would not still be an occasional feature of the Wiltshire, Berkshire and Hampshire countryside. Thatch was also used in various parts of Wessex for the protection of yard and farm walls, which were constructed many years ago of a straw and clay mixture, rather than chalk and flints. The type of thatch employed for wall protection was not usually of the best quality, as used in roof construction. However, the thatchers or sometimes the farm workers, always gave the wall thatch a very neat, attractive, finished appearance.

Thatched cottages can be found scattered throughout the whole United Kingdom, but English counties lying south of Yorkshire and Lancashire possess the greatest number of thatched dwellings. There are today an estimated fifty thousand thatched dwellings in England. However, certain individual counties have a higher proportion than others. In particular, the counties of East Anglia possess a high number of thatched properties, no doubt due to the ready availability of the locally grown Norfolk water reed, much used as a thatching material. The West Country also claims a high proportion of thatched buildings: it is estimated that Devon has approximately eight thousand thatched homes and the smaller county of Dorset four thousand.

The West Country boasts an exceptional number because it has always been largely agricultural in nature and a producer of corn from which straw and wheat reed can be derived for thatching. Other counties of England which have traditionally produced large quantities of corn, also have high proportions of thatched buildings. These counties are predominantly found in the south of England and the Midlands. The term 'corn' covers the crops wheat, oats, barley and rye. The north of England has a relatively small number of thatched houses because of its lack of large-scale corn production. For example, it is estimated that Yorkshire now possesses less than fifty thatched homes.

The type of people living in thatched properties has changed over the centuries. With the more wealthy, the position was largely dictated by whether thatch was in or out of fashion at a particular time. In the past, the vast majority of thatched properties in the countryside were usually occupied by either farm workers, farmers or yeomen but in recent years there has been a general decline in the presence of the traditional ploughmen and other agricultural workers living in the old thatched cottages, as mechanization has made the agricultural industry much less labour-intensive but at the same time more productive. The population has now changed due to the influx into the countryside of a new type of owner. Many thatched dwellings have now been purchased, extensively renovated and modernized by previous city- and town-dwellers, who buy the country cottages as second homes for use at weekends or holidays, with a view perhaps of one day retiring to them. (Property developers have played a role in restoring and modernizing such properties before placing them on the market.) Also, due to the speed of modern transport, many owners now commute each day from their country cottages to lucrative jobs in cities and towns. Other owners are those who have become disenchanted with city and town life and prefer to live in the countryside. The majority of these new-type owners are fortunate enough to be able to afford the additional expenditure required to maintain their properties in good order. The dwellings have thus become more readily resaleable than they used to be and therefore more likely to attract a favourable mortgage decision, if one is required by the new purchaser.

The migration of the new type of person to live in the countryside has also changed the nature of the old village concept. While many of the former inhabitants and especially the younger members have drifted away from agriculture to more financially rewarding occupations and now live in other, more industrialized regions, their replacement by the previous city- or town-dwellers results in the formation of a new mixed community with the remaining agricultural workers and farmers. However, English villages still retain much of their original charm, and thatched cottages remain a distinguishing feature. At the present time, it is true to say that the aesthetic appeal and the charm of a thatched roof are still irresistible to a great number of people. Due to this, thatch can be found covering not only some of the more humble cottages in rural England but also some of the more expensive and exclusive homes. Most of these homes are several centuries old, but even a few newly-built houses are now being roofed with thatch, in an attempt to recapture the styles of England's heritage.

(2)
Thatch Materials

Most old houses which are thatched were originally covered by materials grown in the surrounding areas, and, on economic grounds, this practice is still largely appropriate today. For example, *Calluna vulgaris*, commonly known as heather or ling, is still on rare occasions seen in Scotland on the roofs of crofters' cottages and outhouses. (However, its main present-day use is restricted to the roofs of pavilions and summer-houses.) Ling was originally employed because it grows to a considerable size in the Highlands and yields a readily available thatching material. The plant forms very tough wiry stems on a bushy base, and for thatching purposes the ling heather is cut into lengths of about three feet. The heather is harvested during the autumn.

Broom was once used as an alternative material. This is also a plant which has stiff, long wiry branches. It grows to a height of between three and five feet and is found on dry heathlands and open sunny downs throughout the United Kingdom. It was thus a widely distributed material for the thatcher. Broom is sometimes confused with furze, which grows in the same areas and has similar yellow flowers. The main difference between the two is that broom possesses small leaves while furze has sharp spikes.

Another material freely available for thatching was turf, which was frequently cut and used either alone or together with any natural water reeds which were found growing locally. The turfs were normally cut in lengths of about twenty feet, widths of about two feet, with a thickness of about two or three inches. The turfs were preferably cut from ground which had never been ploughed, in order to ensure that the grass roots had not been severed and therefore were still well entangled. The area of the individual turfs was kept as large as could be conveniently handled. This decreased the total number of joins it was necessary to make between the turfs, when they were laid and sewn on the roof purlins. The turfs were always fixed to the roof with the grass side upwards facing the weather. The root side was therefore in contact with the under roof. Turf was particularly used in Ireland for thatching, but it was also utilized in England. In fact, it was still employed until the beginning of this century for the finishing of reed-thatched roofs, especially in the border counties of England and in Wales. The turfs were used in this application to make a water-tight seal for the top of the roof, known as the ridge area, by fixing them to cover approximately two or three feet down on either side of the apex of the roof.

A much rarer form of thatching employed in England was that utilizing wood chips as the thatch material. The chips were made from hazel wood, in lengths of about two or three feet. The lengths of wood were slightly tapered, so that they became gradually thinner throughout their lengths. The individual lengths were bundled together with all the thick ends of the taper gathered at the same end of the bundle. The bundles were next tied to the roof with the thick ends facing downwards on the roof slope, towards the ground. As the roof was covered, the thick ends of the wood chips were pushed upwards, so that they became firmly wedged together. The end result was a tightly packed wood chip thatched roof. Other fairly uncommon materials which were used for thatching were chestnut shavings and sometimes flax.

A much more common form of thatch material utilized in England was sedge. This is a grasslike plant, found growing under wet conditions in bogs and ditches, and it is still occasionally utilized for the thatching of whole roof areas of houses in the Fens. However, it is now more frequently employed as a ridge material for the tops of roofs constructed from Norfolk reed, which itself is difficult to bend over the apex of the roof because of its toughness. Sedge under favourable soil conditions can grow as high as six feet. It can be cut at any time of the year and is especially pliable and easily bent when in the green condition. With age, it becomes brown and less resilient. It has a life of approximately thirty years when used as a ridge material, and it blends in well with the appearance and the golden brown colour of a Norfolk reed roof.

Most thatched roofs in England are now constructed of either reed or straw. The most common materials are Norfolk reed and combed wheat reed. A material known as long straw is used to construct straw roofs. These natural and cultivated products have been employed for a very long time as thatching materials.

The best known and the most expensive thatching water reed is the *Phragmites communis*. This is the tallest reed found in the United Kingdom, growing up to ten feet in height. However, the reed is normally only suitable for thatching when obtained from certain areas, where semi-cultivation has been carried out to encourage a straight and strong tapering structure. The reeds need wet conditions for growth and they then grow vigorously, forming dense clusters on the edges of fresh or brackish water. The reeds are perennial and spread by the creeping of the root stock. They can be seen in dark purple flower during August and September. They grow on a single main stem. The seed heads, when formed, are brown and feather-like. The leaves are large and pointed, with the appearance of spears. The wet conditions essential for their growth makes an area such as Norfolk, containing a large expanse of lakes, rivers and waterways, ideal for their propagation. Many of these waterways in Norfolk were originally made by the old turf-cutters, when the excavations they left behind became filled with water. Some are fresh water and others brackish. The intensive turf or peat digging took place over many centuries, ranging from Roman times until the end of the fourteenth century.

Without doubt, Norfolk reed is the most famous, tough and durable of thatching materials. The reeds are sometimes called by the alternative name of 'Norfolk spears'. For maximum strength, the reeds should be cut at a particular time of the year, after the frost has killed off the leaves.

Depending on the weather conditions, the best harvesting time is normally between January and March. At this time of the year, flat-bottomed boats loaded high with bundles of reeds can be seen on the Broads. Nowadays, some of the harvest is even exported to the United States of America. The regular cutting of the beds produces reeds which are much superior as regard straightness, taper and strength. The reeds flourish best in standing water, but they have a tendency to build up the soil level as they die back each year and create debris. Regular cutting encourages better growths. Neglected reed beds rapidly deteriorate, and the reeds yielded are no longer of ideal quality for thatching purposes. Reeds cut by hand are often considered superior in strength to machine-cut ones. This is because the former method is more gentle and therefore the material is subjected to less stress than when machine-cut. However, modern machine development has overcome this problem to a great extent. It is now possible to obtain machines which cut and bind a very large number of reed bundles per minute. This innovation has saved the men, who were formerly employed for manual cutting, from the extreme hardship of often wading and working in ice-cold water during the winter months.

As would be expected, many examples of Norfolk reed roofs can be seen in East Anglia. However, the *Phragmites communis* reed is found in many parts of the United Kingdom besides Norfolk, though it is only called by the name 'Norfolk reed' if it originates from that area. A little *Phragmites communis* reed is established in Dorset, in the Radipole and Abbotsbury areas. The use of Abbotsbury reed was once quite widespread in the south of England. The *Phragmites communis* is also found in coastal areas of Wales, Hampshire and many other marshy regions on the east coast of England, such as Suffolk. All these areas have the correct environmental conditions for its growth. However, as in the case of Norfolk reed, the reed beds must be semi-cultivated to produce the high quality reed suitable for thatching. Many of these local marsh reeds are used by thatchers in the vicinities of where they grow and are called by their local names, such as Hampshire or Dorset marsh reeds. Sometimes, even the local town name is used, as, for example, in the case of Weymouth marsh reed and Abbotsbury reed.

Other types of water reeds are frequently found growing alongside the *Phragmites communis*. Typical of these are the bulrushes *Typha latifolia*, also known as the 'Cat's tail' and the 'Great reedmace'. This plant grows up to eight feet in height and flowers in July. It can spread very rapidly. Another intruder in marshlands is the *Iris pseudacorus* or 'Yellow flag' which normally grows to a height of about four feet and flowers in June and July. These varieties of marsh reed are occasionally used blended in minor proportions with the *Phragmites communis* for thatching.

The main alternatives to Norfolk and other marsh water reeds are combed wheat reed and long straw. Both of these nowadays are usually obtained from wheat. However, it is not possible to harvest the wheat intended for thatching with the use of a combine harvester, because it carries out the function of threshing the corn at the same time as it reaps it. The combine then pours the grain into sacks and scatters the straw stalks, which are usually broken into short pieces, on the ground. The advent of the combine harvester, although greatly improving the world's grain harvest, has at the same time greatly reduced the amount of good length straw which is available for thatching. In order to obtain long unbroken straw stalks suitable for thatching, it is essential to harvest the wheat separately, using a reaper and binder. This machine does not carry out a threshing operation as it reaps but just cuts the wheat and binds it into bundles.

The short supply situation of suitable length wheat straw for thatching has been further aggravated by the improved varieties of corn which are now grown to optimize grain yield. These varieties are shorter and stiffer than those grown in the first half of this century. The new varieties of wheat are preferred by farmers because they can tolerate much higher levels of nitrogenous fertilizers to increase their yields. They are also less likely to become flattened by wind and rain. However, the shorter wheats have made straw suitable for thatching much scarcer. Farmers sometimes grow a limited number of segregated acres of a long wheat variety especially for the thatching trade. The thatchers will buy the crop and make arrangements with the farmer for it to be harvested, in a manner most suitable for their own requirements. Particular attention is paid during the harvesting and later handling of the long wheat, to ensure that the stems are not bent or crushed.

The wheat from which the straw is derived should, ideally, from the thatching point of view, be harvested when still slightly green, as this gives a stronger, more durable straw. The length of straw required for good thatching is in the approximate range of two and a half to three feet. Straws which are allowed to grow naturally have a greater strength than when they are artificially forced in their growth. The use of fertilizers tends to make the straw brittle. Wheat was originally combed by hand to remove the grain and leaves without damaging the stalks. This was a laborious but necessary process to obtain a wheat reed suitable for thatching. Combed wheat reed is now obtained from harvested wheat by the attachment of a special reed-comber to the threshing machine, when the threshing operation is carried out. The sheaves of corn are fed into the machine all the same way round. A comb removes the leaves and grain from the wheat but does not allow the stalks to enter the threshing drum to which the combing machine is attached. The straw is therefore not subjected to mechanical stress in the thresher, and the individual unbroken stalks are obtained from the machine in a parallel form, suitable for tying into bundles with the thicker butt ends together. The combed wheat reed is classified as a reed from the thatcher's point of view and is laid into a roof using a similar technique to Norfolk reed or other marsh water reed. Combed wheat reed is predominantly used in the south-west of England and is sometimes known as 'Devon reed'. Many examples of houses thatched with this material can be seen in Devon, Dorset and Hampshire.

In contrast to combed wheat reed, long straw used as an alternative thatching material has been separated from the wheat grain by itself undergoing a threshing operation. In past times, the wheat was cut by hand and then threshed as gently as possible to remove the grain. In present times, the threshing is still carried out as carefully as possible to minimize damage to the stalks, but the straw delivered from a threshing operation is often in a broken and tangled state. This entails a tedious procedure to straighten it by hand. In further contrast to combed wheat reed, the butt ends of long straw have not been gathered together in an orderly parallel stalk fashion. Therefore, long straw has to be laid into a roof by the thatcher using a different technique from that employed for Norfolk reed and combed wheat reed.

Rye straw was once commonly utilized for long straw thatching as well as wheat straw, but it has become a scarce commodity because few farmers grow it nowadays. The decline in rye production in England started in the 1930s when it decreased by over forty per cent. Production received a slight boost during the Second World War period, when government incentives sometimes made it more profitable for farmers to grow it on lighter, poorer soil than wheat. However, production has steadily declined since that period. Rye straw was an ideal material for thatching, as the straw stalks were strong, long and less brittle than wheat straw. Barley and oat straws were also utilized for long straw thatching, but the stalks were less tough than those of wheat straw.

Long straw, derived from wheat, is used as a thatching material in many of the counties of England where large quantities of corn are grown. It is thus predominantly found in the south of England and the Midlands. In particular, many examples of long straw thatched roofs can be seen in the counties of Cambridge and Bedford. However, properties thatched with combed wheat reed and water reeds are also often found in these areas.

Many different terms have been used to describe and measure harvested quantities of reeds and straws for the thatching trade. For example, bundles of reed or straw are sometimes known by the term 'thraves'. The word 'boultings' is also sometimes used to describe bundles of straw or reed in certain areas. With reeds, the term 'bunch' is often used to describe the bundle, in contrast to the 'yealm' used for long straw. A yealm of long straw is normally reckoned to be approximately sixteen or seventeen inches wide and about five inches thick. A bundle of combed wheat reed is known, particularly in Devon, as a 'nitch' and weighs twenty-eight pounds. Norfolk reed is not calculated by weight; it is measured in 'fathoms'. A fathom of Norfolk reed means a tight bundle possessing a circumference of six feet (one fathom) around the butt end of the bundle. The relatively large bundle is made up by gathering together five smaller tied bunches known as 'bolts'.

With regard to the relative costs of the three main alternative thatching materials, Norfolk reed, combed wheat reed and long straw, it is useless to attempt to quote exact figures. They vary constantly due to inflation and to the quality of the particular crop. In the unlikely event of material costs becoming stable, it is still difficult for a house-owner or prospective purchaser to estimate accurately the total cost of a new thatched roof for his house. This is because different roofs have different shapes, and properties are sited in different positions around the countryside. They may be close by or far away from the source of the thatching material to be used. As a rough guide on basic material costs, it can be estimated that the price of Norfolk reed will be nearly twice that of combed wheat reed. The latter material will again be more expensive than long straw. However, the situation is changing due to the increasing difficulty in obtaining good quality wheat straws for thatching. As mentioned, this is due to the short straw varieties now grown for the combine harvester and the need therefore to grow special crops for thatching purposes. In the future, there may well be a trend to use increasing quantities of water reeds for thatching, as the price differential narrows.

The amount of thatching material required will, of course, depend on the size of the house and the number of windows and other architectural features which might modify the total required. The cost of a thatched roof will be greater, for example, if there are dormer windows to be covered and if there are several valleys in the roof structure. This is predominantly because of the extra labour required. Assuming that approximately a minimum one foot thick layer of thatch will be required and working in basic squares (10 ft by 10 ft or 100 sq.ft) of reed material, a small house may require twelve of these squares, approximately six on each side of the roof. Thatchers normally quote prices in individual squares of a hundred square feet of the roof, as measured on the sloping surface. The price of constructing the eaves below is normally included. Each square would contain, for example, just under one hundred bunches of reed. In weight terms, for the whole roof this would amount to several tons of reed. However, the total quantity will be less if the whole roof is not to be renewed but only a new cover layer of thatch to be placed over certain areas of the old. Another factor, which influences the amount of material, is whether fancy raised ridge work is required as a decorative finish to the top of the roof. The extra material and also the time taken to construct artistically a thick ridge layer adds considerably to the cost.

With regard to the expected comparative lives of roofs made from the three main alternative thatching materials, a well constructed, good quality Norfolk reed roof will be expected to last, under England's fairly temperate weather conditions, for a period of thirty to sixty years. This compares with approximately twenty to forty years for a combed wheat reed roof and ten to twenty years for a long straw roof. With the latter type of roof, the life was even less when barley or oat straw was used but a little longer when rye straw was utilized. It is worth studying these figures carefully in relation to the initial thatch material costs. Although in the short term it is much cheaper to have a new roof constructed of wheat reed or straw, this is not usually true in the long term. It is a matter of personal choice whether to pay twice the price for a material, such as Norfolk reed, which can be expected to last twice as long as an alternative thatching material. With an old roof, it may not be possible to make a choice, as it will be only a practicable economic proposition to use the same material as already on the roof.

The longer life of the Norfolk reed is due to its supremely better mechanical stability, when its surfaces are exposed to the rigorous forces of nature. In practical terms, this means that the tough hollow reeds are less likely to break or be crushed when repeatedly buffeted by strong winds, moderate gales and rain. The reeds therefore retain their original cross-sectional areas for long periods of time. This stability prevents compaction of the thatch, which would eventually result in its biological degradation and rot. Pieces of rotted thatch would be carried away in the wind or by the birds, and the roof would slowly disintegrate.

Straws, in comparison with water reeds, possess more oval-shaped cross-sections and are more likely to become crushed. This can easily be demonstrated by squeezing a piece of wheat straw in the fingers and comparing its relative lack of toughness with that of a Norfolk water reed. However, combed wheat reed has a better mechanical stability than long straw, which has had its robustness further weakened by the threshing operation. An important factor, which also affects the relative lives of roofs in the alternative materials, is the particular method of thatching. Reed roofs, consisting of water reeds or combed wheat reed, are constructed in a different manner from long straw roofs. The reeds have all their butt ends together and can be packed more tightly, without the danger of fractures occurring, during the construction of the roof. The finished reed roofs are therefore more densely packed, the stalks are less likely to fracture, and also more material is present to give additional strength and durability for combating the ravages of the weather and birds. A long straw constructed roof has more of the stems exposed to the weather (because they are more loosely packed) than when the combed wheat or Norfolk reeds are used. These, in contrast, have only the butt ends, or the last inch or two of them, exposed.

However, despite these factors in making roof life comparisons, much depends upon the definition of the word 'life'. Many roofs are never completely renewed but have a long history of patching and covering with fresh materials as the need arises. Some roofs which have been constantly well maintained, have lasted nearly twice as long as the 'life' figures earlier indicated. In a repair covering operation, a complete new layer of thatch material is secured over the roof surface, after any rotted thatch material and moss has been first removed. The old thatched roof is always taken down to such a level that a secure base is ensured for the fixing of the new layer on top.

Other outside independent factors also determine the relative lives of thatched roofs. For example, the nearness of trees, whether the property is in a sheltered or exposed position and whether repairs are carried out immediately. This is especially so when birds have damaged the roof when searching for nesting or feeding materials. However, despite all these influences, the skill of the original thatcher employed is of the greatest significance and importance in obtaining a long life for a roof thatched in a particular material.

An example of the influence of extreme weather conditions shortening the life of a thatched roof can be seen in areas where persistent gale-force winds are encountered. It is then essential to use ropes to hold down the thatched roof. Cottages along the entire west coast of Ireland, the western Highlands, the Scottish islands and the Isle of Man are all exposed to particularly fierce winter gales. In these regions, and especially the western coast of Ireland, the roofs will be found covered with a network of weighted or pegged ropes to withstand the gales. Unfortunately, the ropes drastically decrease the life of the thatched roofs as they interfere with the free flow of the rain-water from them. The thatch therefore has to be constantly renewed, sometimes as frequently as every year.

Items other than the thatch material itself are, of course, essential parts of a permanent thatched roof construction. These additional materials include the various items used to secure the thatch to the roof, such as spars, sways, liggers, cross rods, straw ropes, iron hooks, tarred twine and withies. Thatching spars are usually made from split pieces of hazel. Hazel is the preferred wood because it is readily available in the English countryside. It is soft and easily split but at the same time tough and flexible. Less frequently spars are made from willow. This wood easily splits and is flexible. It can also withstand blows and general compression of the wood. The quality of the wood used for thatch-securing is important to ensure that a long life in the roof is obtained, comparable with that of the thatch material.

Hazel was formerly grown on a large scale in coppices to

provide the type of wood required for the thatching and hurdle-making trades. The wood was always of good quality, and the shrubs were carefully tended. At the present time, coppices are declining in numbers, and those remaining are generally less well tended. Hazel is a spreading shrub by nature and unless satisfactorily coppiced will not produce the straight, pliant branches required. Deer, in particular, often spoil the hazel wood by eating the young shoots, and the small trees then branch out into a bushy form, like a gooseberry bush.

A thatching spar is made by first splitting in half, down its full length, a hazel rod of approximately two inches diameter

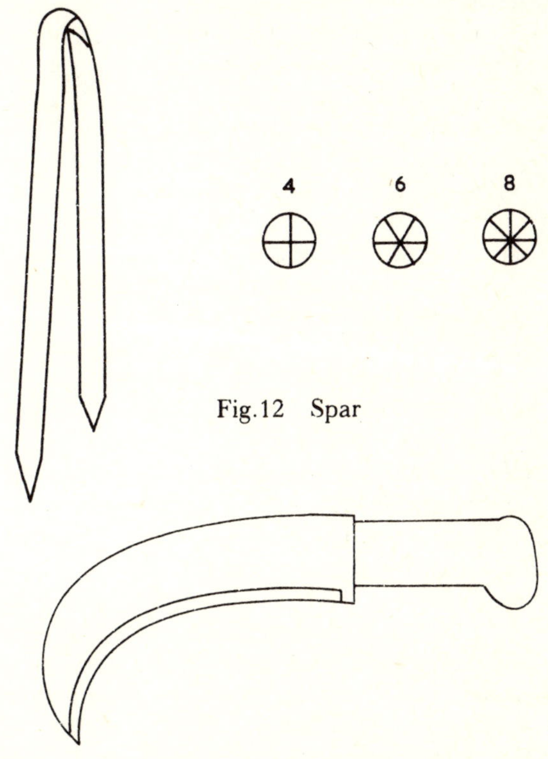

Fig.12 Spar

Fig.13 Spar bill hook

(Fig. 12). The splitting operation is done with a spar bill hook (Fig. 13). It is a skilful procedure to ensure that a true central cut is obtained through the middle of the hazel rod from top to bottom. The two halves are then split further down their lengths, usually into two or three but sometimes even into four. It will thus be seen that four, six or eight pieces of split hazel can be obtained from the original branch. Each will have two flat white sides and one rounded side with the bark still on it. Thatchers either make the spars themselves or obtain them from a supplier or spar-maker. They are normally tied up in bundles of a hundred or more, so that they can be easily transported. Large quantities are used: for example, several hundred spars are needed for the complete construction of a straw roof. The spars are pointed at each end by the thatcher before they are used. He also twists them in the middle, by revolving his hands in reverse directions. The spars are then doubled over into the shape of a staple or hairpin. The twisting action must not break the fibres in the wood, and so the springy nature is still retained. This enables the spars to grip and remain firm and secure when pushed into the thatch material during roof construction. The legs of the spars may be up to two or two and a half feet in length. One leg is often made shorter than the other. Thatching spars are known by a multitude of names in various parts of England. Typical examples are 'brotches', 'roovers', 'scollops', 'sparrods', 'spics', 'spikes', 'splints', 'tangs' and 'withynecks'.

When used, the hairpin-shaped spars are driven into the thatch material to secure the 'liggers', placed over the thatch to secure it down. The liggers are the long lengths of hazel wood which can always be seen on the exterior surfaces of thatched roofs in the region of the ridge and also, in the case of long straw roofs, at the eaves and gable ends. The individual liggers are approximately five feet in length and are usually made by splitting down hazel branches in a similar manner to that used for spar making (Fig. 14). Cross rods are also made of hazel and are often employed as a herring-bone or diamond ornamentation to a thatched roof, when they are secured between two parallel fixed liggers. 'Scuds', twisted ropes of

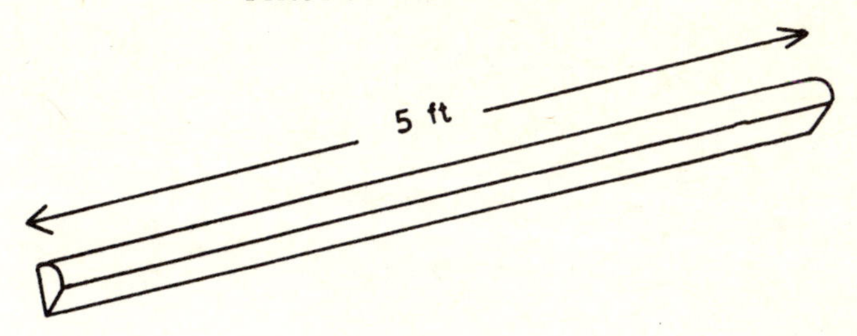

Fig.14 Ligger

straw, are also sometimes employed, in conjunction with spars for the decorative securing of thatch material.

Unlike the liggers, which are always visible on a thatched roof surface, the 'sways' also used for securing the underlying thatch layers are never visible because they are covered by the top surface layer of thatch. A sway is a length of hazel approximately nine feet long with a diameter of about three quarters of an inch (Fig.15). It is used for holding down all types of thatch. The sway, placed horizontally across the thatch course, is permanently secured to the rafters beneath by the use of iron hooks (Fig.16). Iron hooks vary in sizes and lengths, but they all consist basically of a pointed end which enables the hook to be readily driven into a wooden rafter.

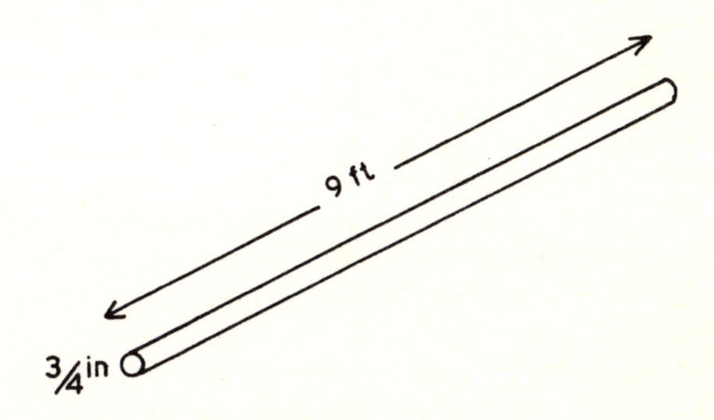

Fig.15 Sway

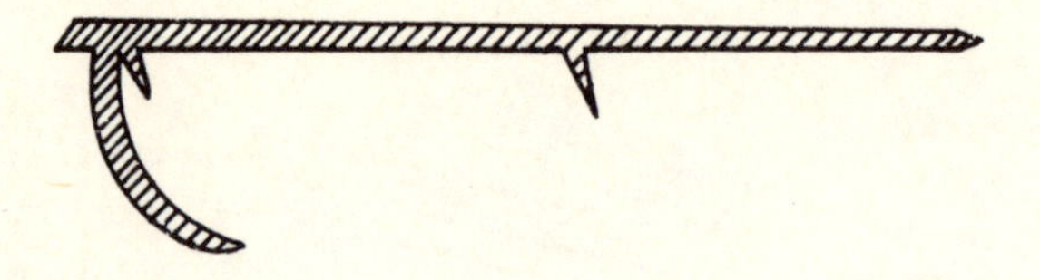

Fig.16 Iron hook

They also have a hook end which can grip over the wooden sway and thereby hold it firmly against the thatch bundle underneath. (Details of the use of sways, liggers and hooks for securing thatch during roof construction are given in the next chapter.)

Thatch material is also sometimes sewn to the roof battens with the use of tarred twine, normally composed of jute or sisal, as these are natural strong, coarse-textile fibres ideally suited for the work. The twine is tarred to make it water-repellent and therefore rot resistant. This ensures that a long life can be achieved when the twine becomes an integral part of the roof construction. At one time, withies were much used for tying bundles of thatching straw to the roof rafters. This was particularly so in Somerset. The withies were obtained from the young shoots of pollarded willow trees which had been originally planted on a large scale, in the ditches of Sedgemoor, during the nineteenth century. The withies were tough and flexible and were used not only for binding thatch bundles but also for the basket-weaving trade. Pollarded willows had their tops cut to encourage the growth of a close rounded head of young shooting branches. Osiers are similar to withies but are cut from particular species of willow. Osiers are also whippy stemmed and utilized for basket-making and wicker work, as well as thatching purposes. In the latter application, osiers were sometimes threaded through and over thatch bundles during the roof construction process. The osiers were then lashed to the roof purlins or rafters with tarred twine.

A final material sometimes employed on the surface of a thatched roof is wire netting. The metal net, of three quarters of an inch mesh size, is always galvanized to prevent rusting and to lengthen its overall life. The wire is subjected to particularly exposed conditions on the roof and the presence of much water. It is also subjected, on occasions, to the smoke from chimneys which tends to yield corrosive products. The wire net may cover the complete area of the roof or it may be placed on certain strategic areas. The wire netting is not an essential part of the roof construction, and it is utilized solely for the protection of the thatch against bird attack. Some recent experiments have been conducted on the use of netting made from synthetic man-made fibres rather than metal. This material has the advantage of possessing complete resistance to corrosive attack. The netting can also be made to look fairly inconspicuous on the roof and therefore more attractive than wire netting.

(3)
The Craft of Thatching

Thatched roofs are designed to shed snow and rain fast. To assist the process, the angle or pitch of a thatched roof should always be considerably steeper than that of a conventionally tiled or slated roof. A pitch angle of about fifty degrees is normally employed with a thatched roof, which means that there is a rise of sixteen inches to each span of twelve inches (Fig.17). The pitch angle of fifty degrees compares with the forty-five degrees commonly employed with a tiled roof and the thirty degrees of a slated roof. As a general rule, the smaller the individual components used in the construction of a roof covering and the more irregular their shape and the more absorbent they are, then the greater is the normal pitch angle employed. There are, of course, many exceptions to this

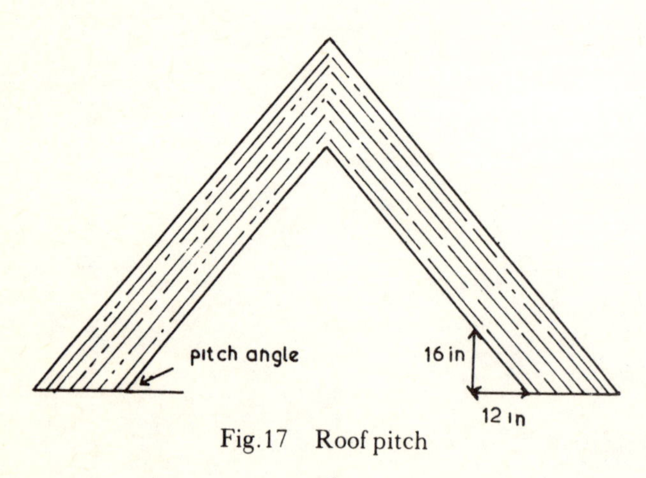

Fig.17 Roof pitch

generalization, and slated roofs may sometimes be seen with equivalent pitch angles to thatched or tiled roofs.

The individual reeds or straws of a thatched roof are open-ended hollow tubes filled with air. The roof is thus relatively lightweight in comparison with its thickness or with one constructed of most alternative roofing materials. The rafters in most old thatched cottages are therefore not usually robust enough to take the weight of heavy tiles. Additionally, an examination from the loft area of the old rafters in a thatched roof will reveal that they are much more widely spaced than those in a tiled roof and their finish is also much rougher. Sometimes the roof purlins were originally made from ash poles directly cut from a copse and then fixed across clefted oak rafters. The basic wooden construction was therefore fairly light, but the strength was quite adequate to support a thatch cover. In some very old cottages, it may be possible even to see the original wattle work built to support the thatched roof secured on top. In a few instances, purlins or roof battens may not have been used but instead a thick mat layer, woven from straw or reed, employed for supporting and attaching the thatch bundles when they were laid.

It is for the main reasons of pitch angle and weight that it is

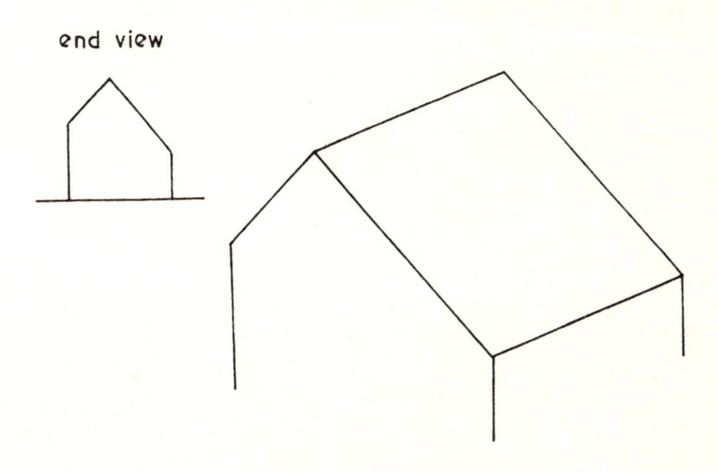

Fig.18 Outshot roof

not usually practicable to have a tiled or slated roof replaced by a thatched one or, conversely, a thatched roof changed to another type. It can be done, but the operation usually involves a great deal of expense, and the timbers would probably have to be changed or altered. The replacement of thatch by slates may also involve the raising of the walls to reduce the pitch angle. Unfortunately, in extreme cases it has even been known for thatched roofs to be covered with corrugated iron in order to avoid further expenditure on the roof.

A great advantage of thatch material is that it can be laid to fit the contours of any shape of roof. However, to be a practicable proposition, the pitch must be steep. In addition to the conventional gabled, hipped and Sussex hipped types, there are many other variations of roof design which can be satisfactorily thatched. The outshot type of roof (Fig.18) is often seen thatched, and in this design one side of the roof extends to below the bottom level of the other side of the roof. The catslide (Fig.19) is also commonly observed thatched, and, with this type, one section of the roof extends below an adjacent section on the same side of the roof. Lean-to roofs (Fig.20) are sometimes thatched for appearance sake when they form a small extension to an existing thatched house. Winged and E-type roofs (Fig.21) are also thatched, and in

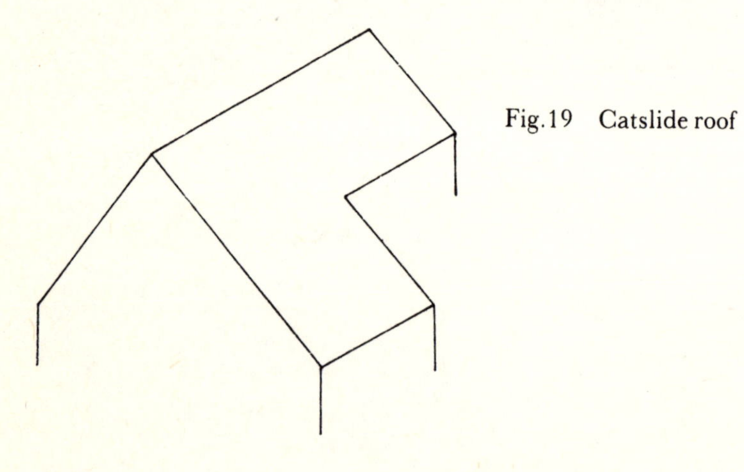

Fig.19 Catslide roof

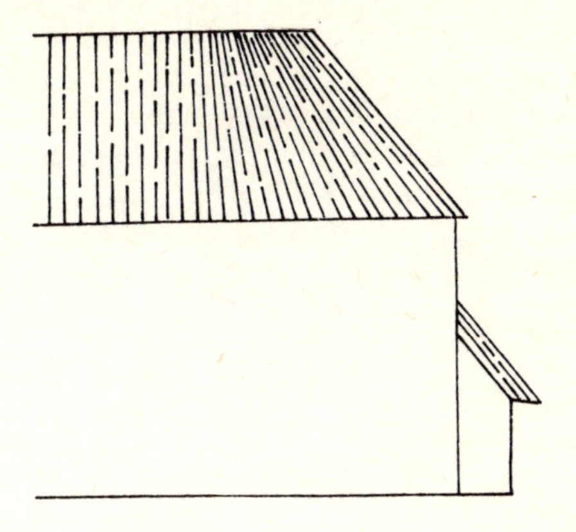

Fig.20 Lean-to roof

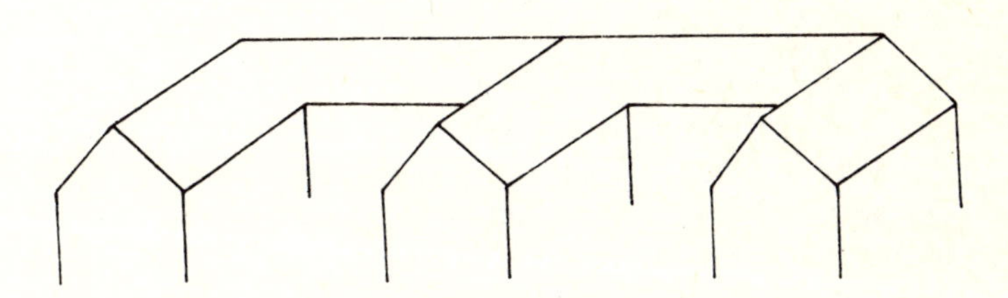

Fig.21 Winged roof

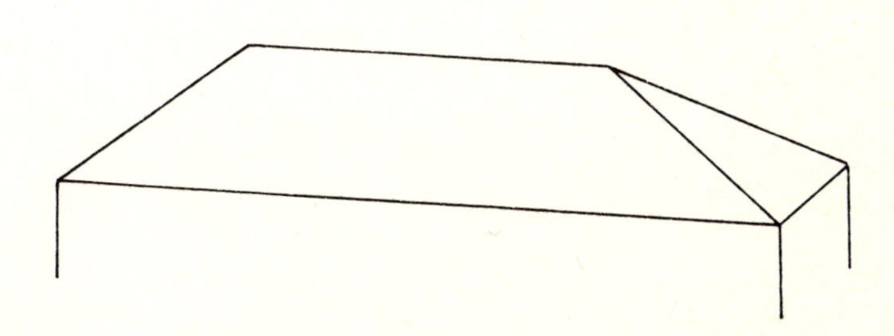

Fig.22 Queen Anne roof

these types the cross wing meets the main roof structure at right angles.

Impracticable to thatch satisfactorily, because of the pitch aspect, is the Queen Anne type of roof (Fig.22), which is low angled. The Mansard roof (Fig.23) presents another

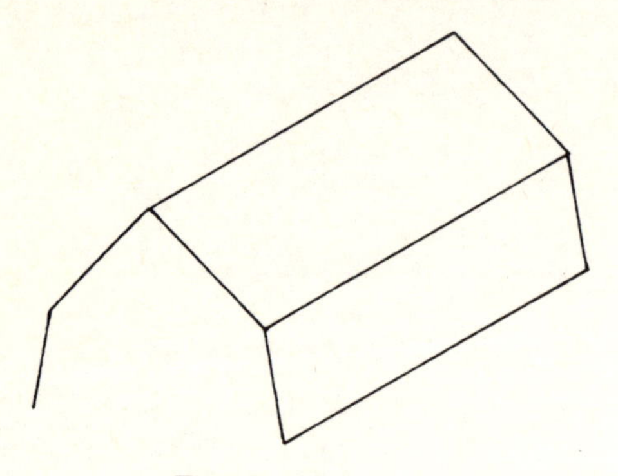

Fig.23 Mansard roof

impracticable proposition because of the particular arrangement of the double slope on each side of the roof. The bottom slope is steeper than the top slope. With a thatched roof, the section at the top or ridge area should be the steepest part of the roof to ensure that rapid water-shedding properties are achieved. The Gambrel roof (Fig.24) would present difficulty and is unsuitable for thatching because of the reduced inclination of the end roof section.

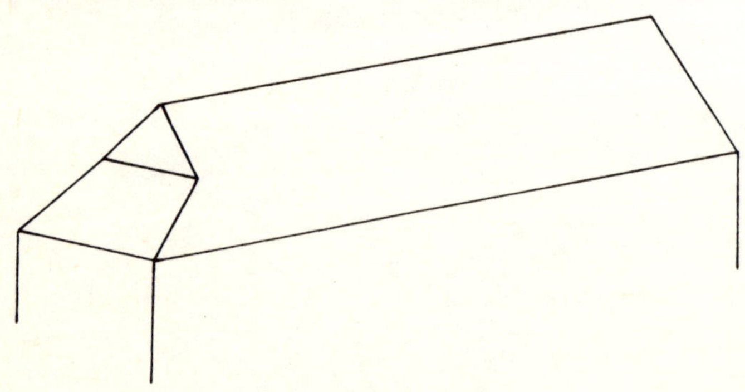

Fig.24 Gambrel roof

A finished thatched roof, whatever its particular shape, always possesses beautiful smooth contours. There are no sharp corners, and even roof valleys are given a swept appearance. The thickness of the thatch layer depends on the quality desired of the roof and the particular pitch. The thickness may vary between nine and sixteen inches. The thicker the thatch layer, the more expensive it will be. The eaves of a thatched roof are always artistically constructed with a good overhang, so that the water shed from the roof is kept well clear of the walls (Fig.25). The eaves normally

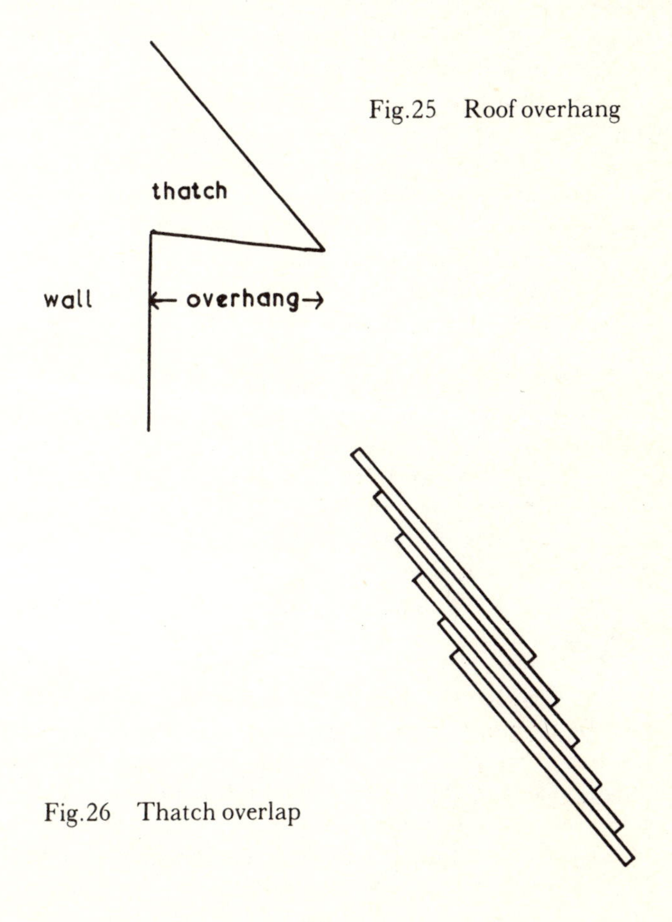

Fig.25 Roof overhang

Fig.26 Thatch overlap

project a distance in the region of eighteen inches to two feet from the walls.

The craft of thatching entails that the reeds or straws in the roof are arranged in thick overlapping layers in a parallel fashion pointing downwards with the slope of the roof (Fig.26). Water droplets can then rapidly cascade over the outside surfaces of the straight individual reeds or straws. The drops of water finally reach and run off the eaves of the roof to the ground. Water should never penetrate below about the top inch of the thatch material surface. In order to ensure this, it is very important that no bent straws or reeds are used in the length of the thatch material, as these would be liable to take water into the body of the thatched roof.

Some preliminary treatment on the ground is necessary on long straw to make it more flexible and resilient before it can be used as a thatch material on the roof. The damping allows bent straws to be straightened, without causing fractures of the stalks. The ground treatment consists of slightly, but uniformly, wetting the straw with water. This is normally done by damping the layers of straw as they are placed on top of one another. The prepared bed is left to soak for several hours. The straw layers are then thoroughly but carefully turned and mixed together to ensure that the same degree of wetting is present throughout the pile and equally distributed over the entire surface area of the straw. This treatment is known as 'yealming', and the tight wet bundles of straw, when they are gathered together for thatching, are known as 'yealms'. The ends of the individual straws in the yealms are made as level as possible.

Far less wetting is needed to make combed wheat reed into a material suitable for thatching. Water is normally just dripped into the ends of the bunches of wheat reed, when they are stacked vertically after their straw butt ends have been levelled by tapping the bunches down on a flat board surface. The wetted bunches are then placed horizontally on the ground and left to steep for a short while before use. Norfolk reed requires no preliminary damping treatment. The reeds are simply dressed by the thatcher on a dressing-board to level the butt ends. The tops of the reeds may need trimming by cutting off any feathery tips still present.

The thatch material has, of course, to be carried up a ladder to the roof before any construction work can commence. This is sometimes done by carrying up the thatch bunches under the arm. However, several yealms or bunches are often brought up at the same time, by packing them in a special carrier which can be borne on the shoulder. A simple carrier may consist of a piece of an ash tree branch, selected in the shape of a two-pronged pitchfork (Fig.27). After the bunches of thatch material have been placed in the fork, a loop of rope is passed over the top of the two prongs. This operation can be readily achieved due to the ease of compression of the ash prongs, which allows the loop to be easily slipped over. The natural springiness of the prongs allows the thatch bundles to be held gently but firmly in the compressed prongs. The carrier, when aloft, may have an adaptation to allow it to be hooked to the roof, so that it cannot slip and the thatcher can work unimpeded. Sometimes, the bunches of thatch material may be transferred to another holder already installed on the roof. This type of holder would probably also have been made by the thatcher out of ash or hazel wood. It would be large enough to hold and retain several bunches of thatch material.

A three-pronged reed fork may also be used for the lifting of reed bundles from the ground to the roof (Fig.28). This type of fork possesses two long prongs and one short prong. The same type of fork is often utilized for the careful handling of reed or straw material on the ground, so that the stalks are not damaged. When the thatch material has been transferred to the roof, it may sometimes be held near at hand to the

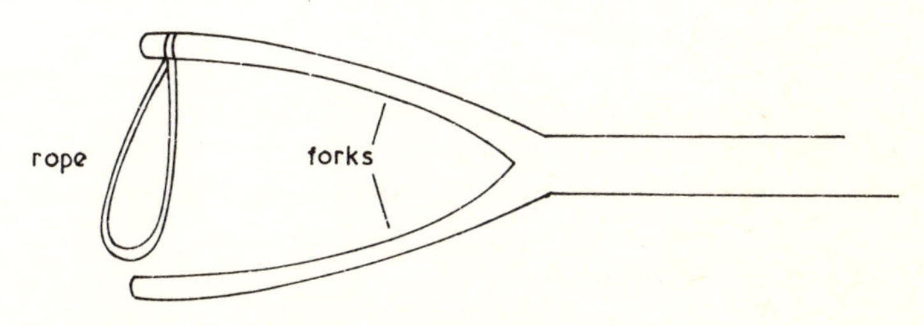

Fig.27 Thatch carrier

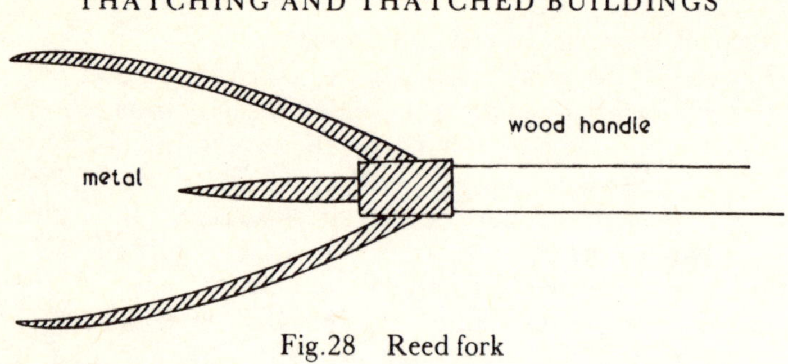

Fig.28 Reed fork

thatcher by the use of a thatcher's spear (Fig.29). The total overall length of this implement is around four feet, and the metal spear section is fitted to a wooden handle. The spear is used to retain the thatch bundles, until they are required, by spearing them to the roof while the thatching operation is in progress.

It will be obvious that during roof construction the thatch material must be tightly secured to the roof, so that it will be able to withstand the future ravages of the weather. For this purpose thatchers employ many different methods, but basically they all consist of combinations of the use of various fixing devices made mainly of hazel wood. Some of these devices are nailed on the top of the thatch to the underlying rafters by the use of iron hooks. Sometimes the hazel lengths may be used, in conjunction with spars, to secure a top coat of thatch to a thatch undercoat already sewn to the roof timbers. At one time, the first layer of thatch material was commonly tied to the roof purlins with such materials as split brambles, straw ropes or withies, but nowadays it is more usual for tarred twine sewing to be used, if this particular technique is favoured by the thatcher.

In the case of long straw and combed wheat reed thatching, it is the usual practice for the thatcher first to spread and also sometimes tie straw bundles horizontally to the roof purlins

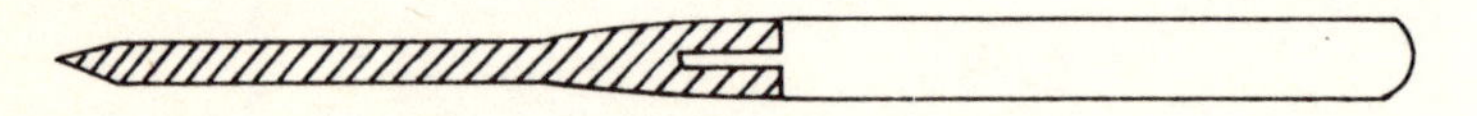

Fig.29 Thatcher's spear

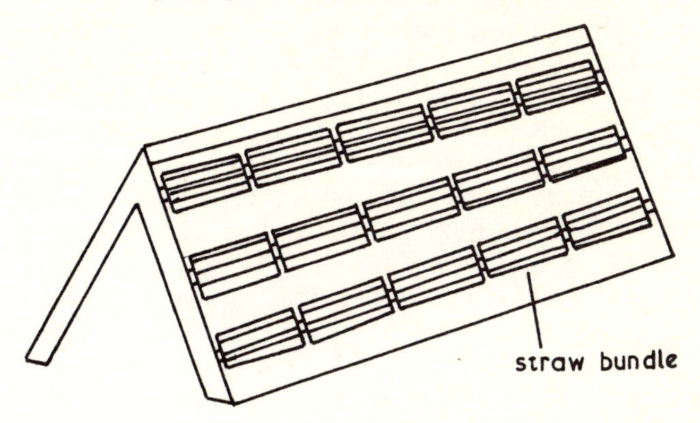

Fig.30 Roof purlins

before laying the actual thatch layers (Fig.30). This procedure eventually gives a neat appearance to the inside of the finished thatched roof when viewed from the loft area. This preliminary operation is not done with Norfolk reed, which is always attached directly to the rafters without the pre-laid underlying layer. However, with Norfolk reed it is the usual practice for the thatcher to push in a back filling of reed behind the reed bunches, after several of them have been laid and fixed. This prevents the ends of individual reeds penetrating into the loft area when the roof has been completed. It also helps increase the tension on the fixed reed bunches after they have been fixed to the roof.

After any preparatory work has been carried out, the thatcher normally starts to construct a new roof by first fixing a large thick yealm or bunch of thatch material to the angle formed between the gable and the eaves at the bottom right hand side of the roof (Fig.31). Horizontal layers, called 'courses', are then gradually built along from the corner, together with vertical layers, called 'lanes', which eventually extend from the eaves to the ridge of the roof. Lanes are normally in widths approximately two and a half feet. In each completed lane, there is no appreciable weight to hold the thatch down other than the overlapping thatch material from the course laid above. An extra yealm or bunch is normally used at each bottom corner of the roof to obtain additiona strength and thickness and eventually to achieve better weather-proofing.

The thatch material is securely fixed along its full width, as it is laid by a variety of fixing techniques which depend on the actual material being used and the individual preference of the thatcher. A common method, particularly with Norfolk reed, is to place a length of wood, called a 'sway' (as described earlier), across the top halves of several thatch bunches as they are laid side by side on the roof. The individual bunches are normally temporarily held in position with the use of iron hooks, until the sway can be permanently fixed across them by driving iron hooks into the rafters. The exact position of the rafters is found by probing with a needle. The top parts of the hooks grip the sway and hold it firmly down on the thatch bunches which are thus securely fixed to the roof. The iron hooks which were used temporarily to hold the thatch bunches in position are then removed. The top coat of thatch is next secured to the thatch undercoat with spars.

It has been mentioned that, in an alternative method, the

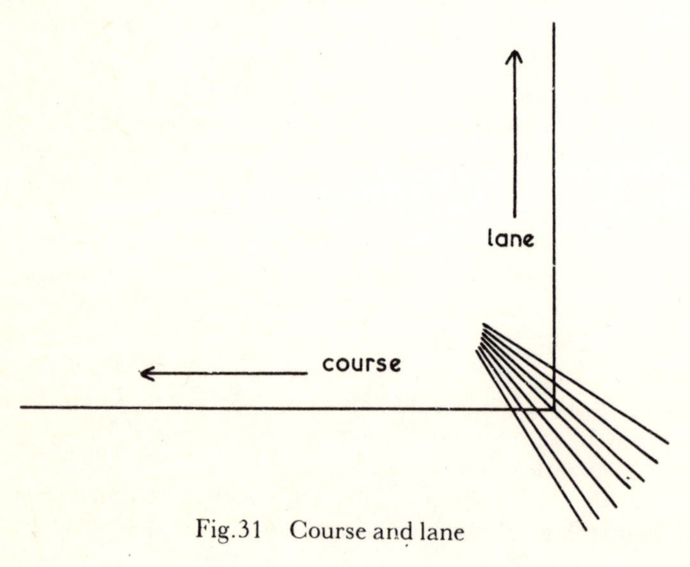

Fig.31 Course and lane

undercoat material is sometimes sewn to the roof battens with tarred twine. This is a particularly common method used in the construction of combed wheat reed and long straw roofs. During this procedure, it is usual for the thatcher to have an assistant in the loft area of the house, who can push the needle backwards through the thatch material as the sewing operation progresses. During this process, the twine is looped around the thatch bunch and the supporting roof batten underneath. The twine is pulled tightly when the stitch loop has been completed.

As the layers of thatch are built up in a lane, from the bottom to the top of the roof, they are placed and secured to overlap considerably the previously-laid ones. The layers on top therefore eventually hide from view the sways, spars and other fixing devices on the thatch layers underneath. This means that the sways are never visible on the exterior surface of the finished roof. The exception would, in theory, be the very top course on the roof where the sways would remain exposed, because there would not be a further thatch course to cover them. However, in practice these are also eventually

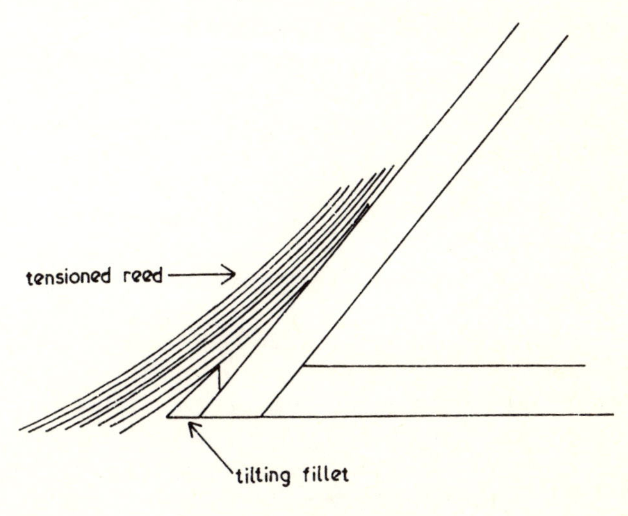

Fig.32 Tilting fillet

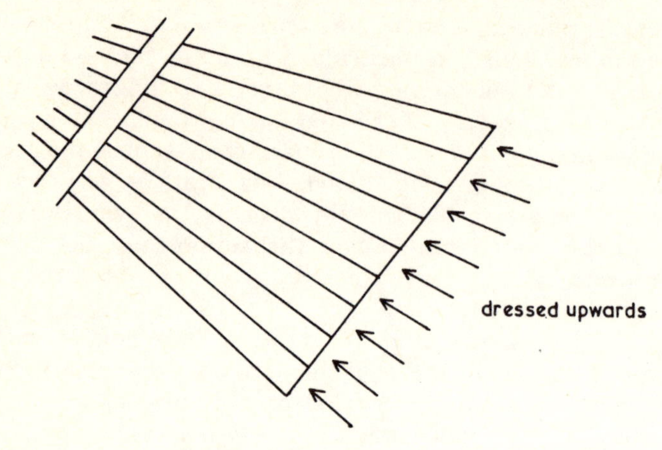

Fig.33 Bunch tightening

covered, as it is in this region that the ridge is constructed. As each lane of thatch material is completed, the thatcher moves his ladder to the left to start the next lane. Sometimes thatchers lay two lanes at a time. It is the practice for thatchers always to work from right to left along the roof. As the individual horizontal courses are progressively increased in length, temporary needles are thrust in at the ends of the courses to keep the thatch material in the courses tightly packed and square.

In the case of reed laying, a tilting fillet has usually been placed at the eaves level, so that the first course laid will be tensioned at an angle to the rafters (Fig.32). The technique helps to ensure that the final exposed surface of the roof will predominantly consist only of the butt ends of the reeds. As each bunch of reed is laid, it is worked upwards as it is fixed, so that the individual reeds become slightly staggered as the face is tightened. The larger butt ends of the reeds are always at the lower end of the thatch bunch when it is placed on the roof, and the slimmer ends are always at the top. The continued working of the reeds upwards, tightens and packs them close together (Fig.33). This procedure ensures that only the last inch or so of the butts will eventually remain exposed to the weather.

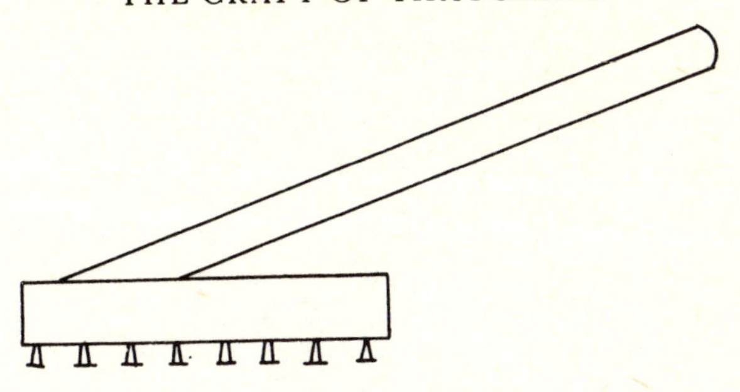

Fig.34 Leggett

The tool used by the thatcher for this reed-dressing operation is commonly called a 'leggett' (Fig.34). This tool has the appearance of an oblong wooden block fitted with a handle. The block is frequently made of poplar wood because it is of a more fibrous nature and less likely to split than other woods. The base of the block is either grooved or patterned with a series of raised horseshoe-nail heads, to assist the pushing of the reeds into position.

The precise type of leggett utilized depends on the thatcher and the material. The raised nail variety is popular in Norfolk for Norfolk reed, while the diagonal wooden groove type is preferred in the West Country for combed wheat reed. The

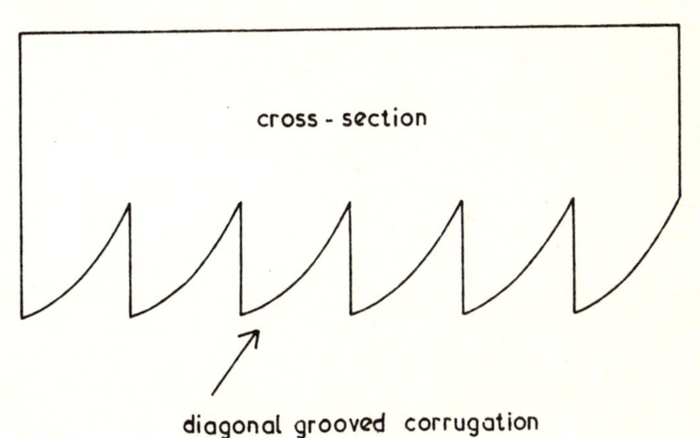

Fig.35 Leggett base

type of wood groove arrangement is usually in the form of a diagonal corrugation (Fig.35). The grooved type is also commonly known by the alternative name of a 'beetle'. In Dorset, it is also sometimes called a 'beater' or 'pomiard'. In Somerset, the term 'flatter' is frequently encountered. Leggetts vary in size, and some are made quite small for use in restricted corners and awkward places. The small variety leggett is also sometimes called a 'moulding plane'. Most leggetts are fitted with hooks on their back surfaces, so that they can be readily hooked into the thatched roof when not in use. This procedure often saves the leggett from falling to the ground and therefore the need, on occasions, for the thatcher to retrieve it.

Leggetts, like so many tools used by the thatcher, are normally individually selected to his own requirements. In past days, the local blacksmith frequently helped in making tools, which would often be assembled from various odd spare parts derived from other farm implements. Items such as knives, sickles and scythe blades were adapted with different types of handles and then used for the shearing and cutting of

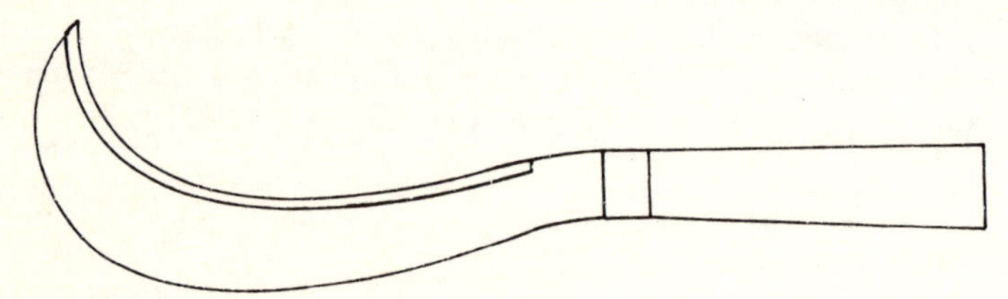

Fig.36 Thatcher's hook

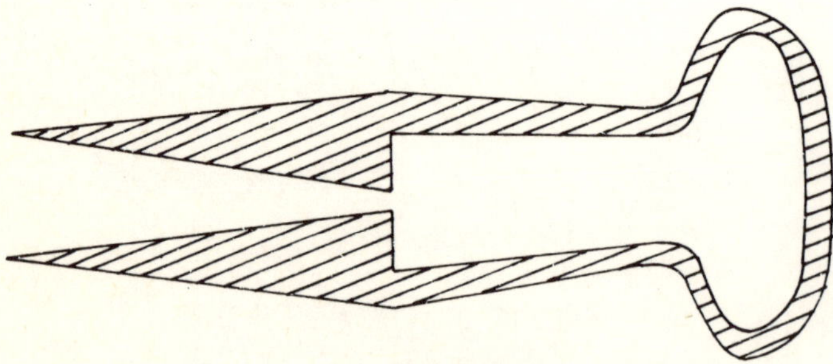

Fig.37 Hand shears

the straight edges of straw-thatched roofs. They were known by the name of 'thatchers' hooks' (Fig.36). Similar tools are now manufactured commercially but are still used for the same purpose. Sheep shearers' clippers are sometimes adapted as hand-shears for the neat trimming of thatch material ends (Fig.37). Many designs of needles are used by thatchers when the thatch material is secured to the roof by sewing. The needles vary considerably in length and size, but most possess a harpoon or spear-shaped head and a large base eye for threading the tarred twine (Fig.38). Rakes used for combing long straw roofs are also frequently hand-made to a particular thatcher's design. A simple thatcher's comb may be made by driving a series of long nails through part of the length of a three-foot strip of wood (Fig.39). The protruding nail points constitute the comb teeth, and the nail-free wood section is used as the handle. Another tool sometimes used by the thatcher is a guillotine, for the trimming of thatching spars, sways and liggers. The wood is placed in the U-shaped recess of the implement, and the blade then squeezed shut (Fig.40). Thatchers also use wooden mallets for tapping spars into position and metal hammers for driving iron hooks into rafters. Most of these tools, with the exception of the leggett, are utilized by the thatcher for the construction of long straw roofs.

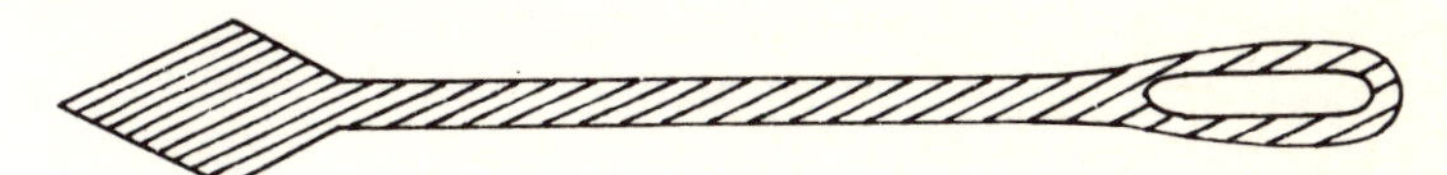

Fig.38 Thatcher's needle

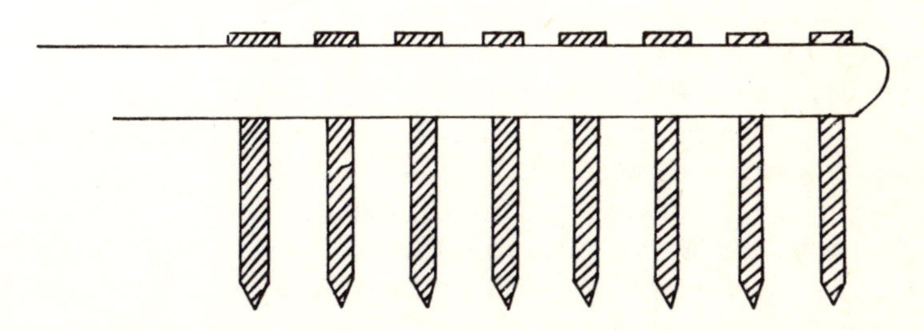

Fig.39 Thatcher's comb

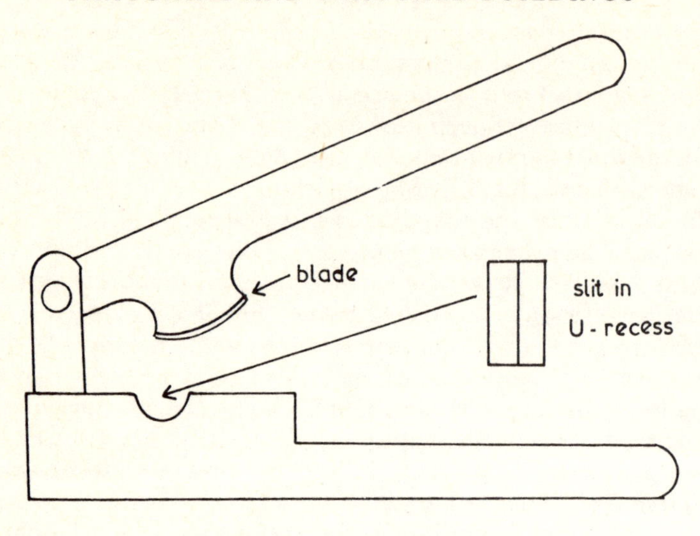

Fig.40 Guillotine

As mentioned earlier, the technique for laying a long straw roof is different from that for a reed roof. This is because the straws in the yealms are in a more random array of sizes, and all the butt ends are not lying together in exactly the same direction. The straws are also not so tough as the reeds, and the long straw is therefore not packed so tightly during roof construction. A long straw roof possesses a much looser appearance than a reed roof. Also, the long straw is not laid at an angle to the roof rafters, so no tilting fillet is normally used. With the long straw type of thatching, as each lane is finished, it is combed downwards and beaten to remove small pieces of straw, to arrange the long straws in the desired direction and to improve the overall appearance. This downwards combing operation is in direct contrast to the upwards dressing and beating with a leggett utilized in reed laying.

During the construction of a course, the long straw yealms are fixed by pushing spars into the undercoat of thatch. Spars are also used to tighten and fix the individual yealms to the ones laid alongside, after any bent straws between the yealms have first been straightened by hand. Spars are also driven in over sways to secure together the courses which overlap one another. It will be apparent that many spars are utilized, at distances of a few inches apart, in the construction of a complete long-straw roof. During the laying operation, the yealms are tightened in such a manner that the ends of the straws tend to face slightly outwards towards the weather. This is done to give the finished thatch extra durability. The roof is then finally raked and given a press down with the back of the rake. After the ridge has been completed, the whole roof area of long straw is usually sheared down with a thatching knife and the eaves clipped. Long straw roofs are always finished with a decorative securing arrangement of spars, liggers and sometimes cross rods around the eaves level and also the gable ends (Fig.41). These securing liggers are always fixed before the gable ends and eaves levels are finally cut by the thatcher. This is necessary in order that a neat edge cut can be obtained because of the looser nature of the long straw thatch.

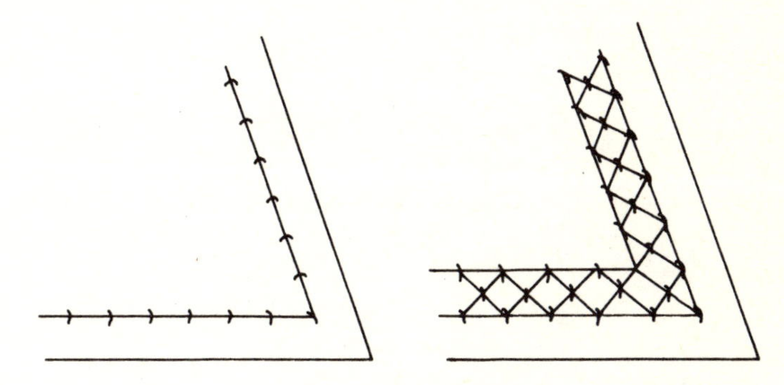

Fig.41 Long straw finish

Combed wheat reed and Norfolk reed roofs are not finished in this manner because they are more rigid. It is thus easy to distinguish a long-straw-constructed roof from a reed roof, due to the additional decorative securing arrangements employed with the former type. Roofs constructed of long straw always have a less stiff and usually a darker, more artificial, loose appearance than those constructed of reed, which appear rigid and possess a texture similar to a short-cropped hard brush.

Combed wheat reed produces a roof of similar texture to Norfolk reed. However, it can readily be distinguished from it as it can be seen on close study that the eaves and gable ends of the thatched roof made with the combed wheat reed, have been cut with a knife or trimmed to the line and form required (Fig.42). This is not the case with a Norfolk reed roof which is not finished by cutting in this manner, the reeds having only been dressed into position to form the desired shape and line. A further distinguishing feature is that Norfolk reed roofs assume a golden-brown colour on ageing, which contrasts with the more greyish colour of an aged combed wheat reed roof.

Considerable roof constructional care is taken with all the types of thatch materials to ensure that waterproof junctions

Fig.42 Cut and uncut reed

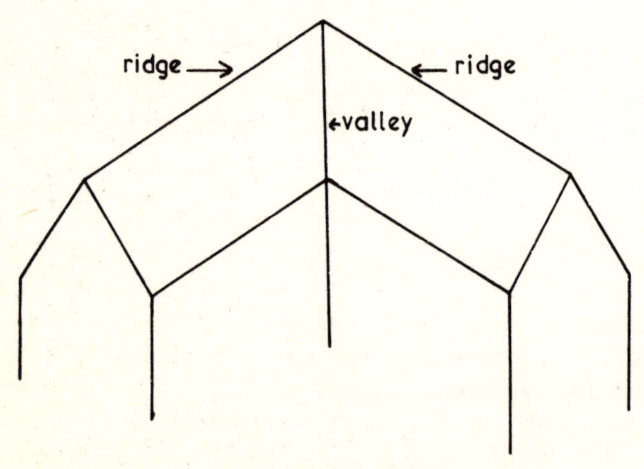

Fig.43 Roof valley

Thatching a cottage at South Uist, Scotland.

Abernodwydd farmhouse, now at the Welsh Folk Museum, St Fagan's near Cardiff.

Above: A thatcher at work at Cockington, near Torquay, South Devon.
Below: Thatching in progress at Woodbastwick, Norfolk.

Above: The Cottage Museum at Great Bardfield, Essex.

Below: Thatched cottages at Birdbrook, Essex.

Above: Thomas Hardy's cottage at Higher Bockhampton, Dorset.
Below: Anne Hathaway's cottage at Shottery, Warwickshire.

Above: An unusual roof level at Selworthy, Somerset.

Below: Buckland-in-the Moor, Dartmoor.

Above: A thatched church at Little Stretton, Shropshire.

Below: Salhouse parish church, Norfolk.

Above: The Congregational Church at Roxton, Bedfordshire.

Below: Possibly the earliest non-conformist chapel in England, at Horningsham, Wiltshire.

A crofter's cottage at Rossaveal on the western coast of Eire

are made at the chimney flashings and also at any roof valleys, where appreciable water flow will occur during rain-storms (Fig.43). The thickness of the thatch layer is increased at valleys, so that the smooth-swept appearance of the thatch is maintained. This also has the effect of spreading the rain water flow over a larger area of thatch surface. A fairly wide concrete ledge is usually built immediately above the chimney flashing, to overlap considerably the thatch material and to cover the chimney thatch junction area.

It is essential that the whole roof area of thatch material is evenly compacted throughout, during construction, so that the possibility of sagging is avoided. It is thus imperative that a good thatch is laid firmly, which is the reason for the large amount of beating or dressing which is required during the thatching operation. If sagging occurs, then rain, over a period of time, will reveal this by gradually channelling a preferred passage for the water from the ridge to the eaves. This particular area of roof will then require immediate repairs or the life of the roof will be much reduced.

The thatch material over windows has to be skilfully dressed, or trimmed and cut, to ensure that the maximum of light will be obtained in the upstairs rooms. This shaping of the thatch has to be done without affecting the overhang which will carry the water well away from the windows and

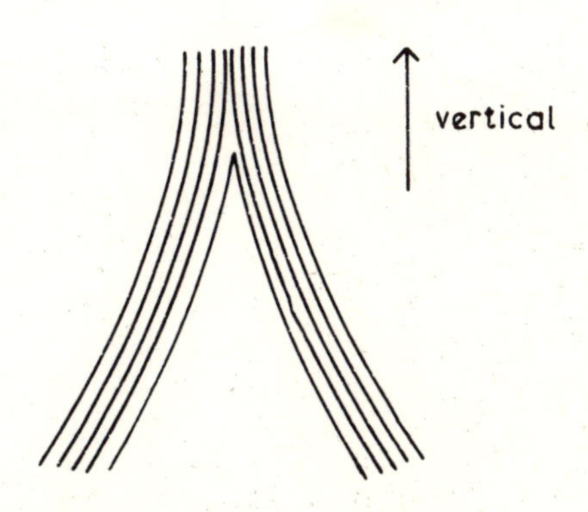

Fig.44 Ridge top cross-section

walls. It must also be done without detracting from the finished appearance of the thatch or interfering with the opening of windows. Dormer windows are always given a swept appearance in the shape of an eyebrow. However, some restriction on the opening of upstairs windows is sometimes unavoidable. Considerable skill is additionally required in obtaining clean, straight edges to the eaves and gable ends of the roof. Gable ends in particular must be exceptionally strongly made because their under thatch sides are usually prone to exposure to wind pressure and lift.

To ensure maximum water-shedding ability, the inclination at the top or ridge of a thatched roof is always very steep indeed, and the thatch material is closely packed into a near vertical position (Fig.44). This near vertical top layer is firmly secured by thatching spars to an underlying horizontal roll of thatch material, which has been previously fixed along the apex or summit of the roof (Fig.45). The underlying roll acts as a secure base for the spars to be pushed into, and it also gives some compressibility below the top course of thatch. During laying, the thatch straw ends at the top are arranged higher than the roof summit, so that they can be later bent over the apex of the roof. Material from each side of the roof is then interlocked and sewn into position with tarred twine on the opposite side of the thatched roof surface. A series of apex rolls of decreasing diameters may be laid progressively over

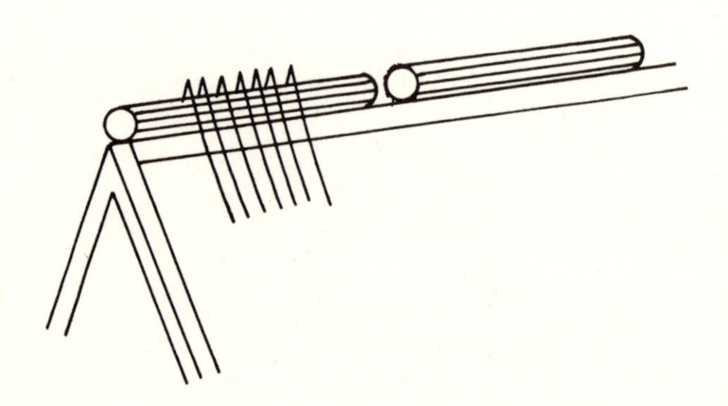

Fig.45 Apex thatch rolls

one another as the ridge construction proceeds. This ensures that the final apex is very narrow.

The ridge work is carried out to finish the top of the roof and to make a completely water-tight seal. The material used for this purpose is generally straw. In the case of Norfolk reed, which is relatively non-flexible, it is the common practice to use sedge. If the sedge has been allowed to become hard and dry before use, it is first softened by damping with water. This process enables the sedge to be readily manipulated and easily bent. The ridge may be constructed at the same surface level as the thatch material on the main roof (Fig.46). A more expensive method involves the building up of a thick ridge layer, up to four inches above the surface level of the main roof surface of thatch material (Fig.47). The bottom edge of this raised ridge may later be cut into various artistic shapes, such as crescent-shaped scallops and points, or decorated in other

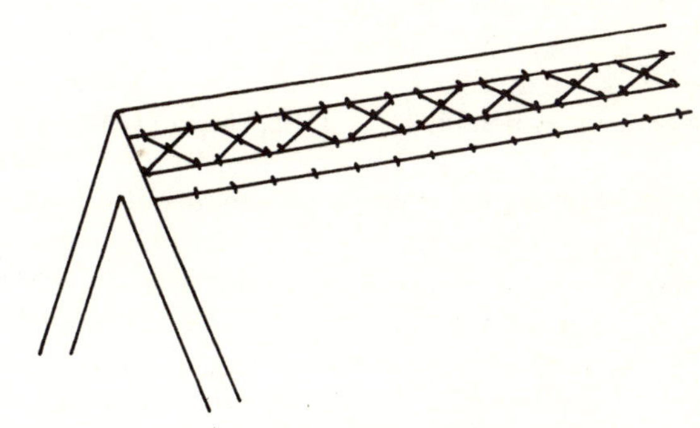

Fig.46 Level ridge

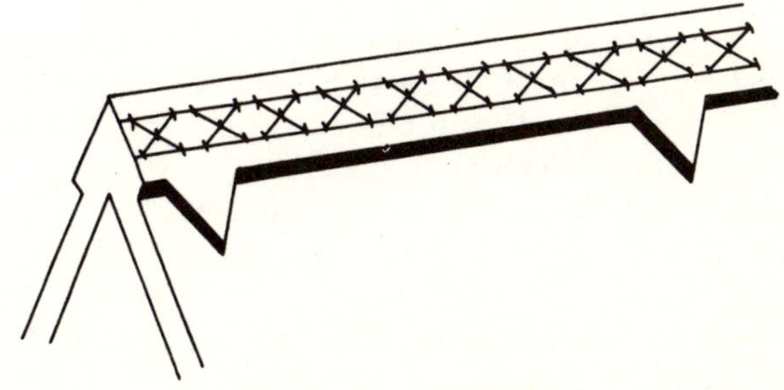

Fig.47 Raised ridge

desired ways. The extra material needed to build up the thick ridge layer and the time taken in its artistic construction make the work expensive. The liggers and spars, used on the exterior surfaces of both types of thatch ridges to secure them, are always visible. The spars are driven into the thatch at distances of approximately a few inches apart over the liggers. They may be arranged in an ornamental fashion with decorative cross-pieces which are also secured by spars, between the liggers. The spars are always driven into the thatch with enough slope to prevent any water being accidentally turned into the thatch when it rains. All finished ridges must be firm and solid. A narrow apex is also an essential characteristic, from both the water-shedding and artistic points of view. During ridge construction, if a chimney stack is in the middle of the roof, then it is sometimes the practice to start at each side of it and work outwards towards the hips or gable ends of the roof.

It is always advisable, after about five to seven years, to have a thatched roof cleaned and patched. The regular inspection and the carrying out of maintenance repairs on a thatched roof will certainly prolong its overall life. In the long term it is also the most economical approach, because if the whole roof is allowed to deteriorate grossly then the only solution may eventually be the complete renewal of the roof. This will be an expensive operation. It may additionally involve the replacement of some of the roof timbers, which may also have rotted if they have been left in contact with damp compacted thatch for a long period of time. Sometimes it will be found that repairs and thatch renewal will be needed more frequently on one side of the roof than the other, because one side of the property may be more exposed to the weather and receive less sunshine. It will become apparent that this side deteriorates more quickly than the other.

During repair work, special attention will be paid by the thatcher to the ridge and also to the eaves and gable ends of the thatched roof. A good ridge is the essential primary requirement for a waterproof roof. When patching or repair work is carried out, the new thatch material is frequently dressed as a layer over the old, after any decayed thatch material and moss have been first removed. The stability of the new thatch layer can only be as good as the strength of the underlying layer, to which it is attached. The new thatch material is secured to the old by the use of thatching spars and these are driven into the underlying thatch to secure the sways placed on top of the new thatch. New isolated thatch patches look yellow, but the appearance soon changes with time to blend with that of the original thatch material which is darker in colour. When bird holes are first noticed in a thatched roof, it is advisable immediately to fill them by pushing straw into the holes, provided, of course, they can be reached without causing other damage to the roof. This prevents birds enlarging the holes still further, before a more permanent repair can be done by a thatcher.

A mesh of wire netting is sometimes used to discourage birds from pulling out straws from thatched roofs. In the case of the long-straw type, the whole roof area may be covered with wire netting, as, due to its looser nature, it is more prone to attack. With combed wheat reed and Norfolk reed, it is normal practice to cover only the ridge, gables and eaves sections of the roof. When netting is fitted, it is usually done in such a way that the joins in the wire mesh lengths and the parts secured by nails at the eaves and gables can be quickly released in an emergency. This is so the whole netting can be more speedily rolled up and removed in the event of a fire.

Some thatchers will never recommend the use of wire netting because it imposes an additional strain on the thatched roof. Strong winds have a slight tendency to lift the thatch material, and the restraining wires then exert a frictional force on the surfaces of the reeds or straws with the possibility of some breakages occurring. The thick broken ends may become trapped under the wire, and they may eventually turn inwards as they cannot escape or be blown away in the wind. The broken trapped straws or reeds can then allow rain water to seep below the top thatch layer, and rotting of the thatch beneath may be initiated.

Even when straw or reed breakages have not occurred, the presence of a wire mesh lying in contact with the thatch surface impedes the flow of rain water as it cascades down the roof. The residence time of water on the roof is therefore slightly increased, and there is a greater chance of water penetration below the top inch of the surface thatch layer. Many people also consider that wire netting spoils the look of the thatched roof. It is thus an individual choice when weighing up the advantages and disadvantages of using wire netting. A thatcher will normally never recommend wire netting if the thatched roof is near trees, as it tends to trap leaves and twigs on the roof. This further aggravates and impedes water flow on the roof. The presence of rotting leaves also encourages the thatch underneath them to rot.

(4)
Thatchers

In the past, thatchers were always local men, and the craft was usually handed down from father to son. Due to the absence of transport, little travel took place outside a particular area: as a result, the thatchers of East Anglia and Devon became particularly renowned for their craftsmanship. Nowadays, with the availability of modern transport, thatchers can travel longer distances. However, most thatchers are still usually local men and originate from the areas in which they work. Many have still learned their trade from their fathers, and the sons have continued to operate the family business. Most thatchers are therefore self-employed and have retained their originality and individual work methods. Many thatchers employ apprentices who are also often local young men who are keen to learn the trade and eventually set up businesses of their own.

Thatchers have always taken pride in their craftsmanship, and, of course, they still do. Besides financial reward, most obtain great satisfaction by artistically working with nature's materials and using old-established techniques. The type of craftsmanship involved cannot be hurried. The renewal of a fairly large house roof may take a skilled thatcher and an apprentice at least five weeks. A similar roof area could probably be tiled in a few days. Even a medium-sized cottage roof may take three weeks to thatch. The slow, artistic nature of the work ensures that master thatchers are much in demand, and it is usually necessary to book their services well in advance. Some thatchers may already be booked with enough work to last them a year ahead. It will therefore never be easy to get a particular thatcher of your choice at short notice.

Thatchers are men who enjoy working in the open air, but thatching is a fairly solitary occupation. Most of the day is spent in isolation on the roof. Even when a thatcher has an apprentice working with him, they are frequently apart. The apprentice may be on the ground preparing the thatch material and carrying it up the ladder. Perhaps he may be in the loft space, assisting in thatch sewing but still a little separated from the thatcher on the roof.

The narrow streets of old villages were not designed for modern traffic. This often produces a hazardous situation for the thatcher, perched on the top of a ladder, when working on a cottage situated in a busy village street. The bottom of the ladder has frequently to be placed precariously in the road, as often there will be no pavement or front garden to the cottage. The thatcher will normally be able to arrange with the local police for a temporary 'Slow' warning notice to be displayed along the roadside, to give him a little protection from the motorist. The ladder may have to stay in the road during the whole of the thatcher's working day and also over a period of several weeks, as the roof construction process cannot be hurried. Operating in a constricted space, such as a village street, also hampers the progress of the work. It is difficult to unload considerable quantities of straw or reed just outside the cottage. Also, each evening the site outside the cottage will have to be tidied, so that no hazards remain during the night for the motorist or the general public.

The craft of thatching is very harsh on the hands, due to the continuous handling of wet straw or manipulating of tough water reeds. Some thatchers wear a special leather glove to protect the palms of their hands during their work on the roof. The palms of the hands are particularly prone to damage, as they are often used to assist the dressing of the thatch material into the desired position. The twisting of spars and the general labour of securing thatch material to the roof are also abrasive to the hands, especially when the task is carried out in the winter. However, severe winter weather greatly curtails a thatcher's work.

The weather plays an important role in thatching and determines the best time for the operation to be carried out.

Rain will cause the thatch material to become too damp, and the work of handling excessively wet thatch makes it difficult and even more unpleasant for the thatcher's hands. Toiling in very cold conditions in the winter is also not suitable because frost has the effect of stiffening straw and making it unworkable. Working in high winds is also avoided, wherever possible, because the thatch material is blown about and generally becomes unmanageable. The ideal conditions for thatching are therefore on windless, dry and warm days. However, thatchers cannot always afford to wait for ideal weather. They have to compromise and work when they consider the weather reasonably tolerable. (In the old days, a thatcher would normally never remove more of the old thatched roof than he knew he could re-cover with new before he finished work for the day. The securing of thick polythene sheets over the roof overcomes the problem nowadays and is also useful in emergencies, when the weather suddenly becomes bad.)

The task of thatching is also hard on the thatcher's knees, due to their constant contact and their inevitable rubbing against the roof and ladder as the thatching work proceeds. Some thatchers protect their knees by strapping on knee pads; in the past, these were normally made of tough leather and were frequently felt-lined to make them more comfortable. Many thatchers also utilize a special ladder to ease the pain on their knees. The two long lengths of the ladder, to which the rungs are secured, are made flat and wide. This avoids the thatcher's having to kneel on a sharp edge, as frequently happens with a conventional ladder, where the surfaces facing the user are normally fairly narrow. The flat and wide lengths of the thatcher's ladder also help to spread its weight over a larger part of the thatch, when the ladder is leaning against it. This stops the thatch being excessively compressed.

The type of thatching material used by a thatcher in a particular region will still largely be determined by its ready local availability. For example, Norfolk reed will be the predominant type in Norfolk, combed wheat reed in Devon and Dorset and long straw often in the Midland counties. However, all the three types are utilized to some extent outside their main regions of origin or traditional use. Thatchers will often specialize in either reed or long straw thatching, depending on the area preference in which they operate. However, despite this specialization, most thatchers are still adept at working with all types of materials. In addition, thatchers sometimes make use of suitable local marsh water reeds which may be cultivated in their areas.

Some thatchers also specialize in the art of ornamental ridge construction. These elaborate, fancy-cut and decorative finishes to a thick raised roof ridge are expensive and time-consuming to complete, but the end products are delightful to the eye. In addition to ridges, raised ornamental thatched apron areas are also often specially constructed below window spaces, when they exist in the main body of the thatched roof surface. These aprons further complement and enhance the overall beauty of the completed roof. Thatchers who do specialize in particular types of thatching sometimes make this apparent in their advertisements in the 'Yellow Pages' classified telephone directory. A fairly comprehensive list of thatchers available in an area are listed under the section headed 'Thatching' in the directory.

Despite the fact that most thatched dwellings are a century or more old and that total numbers are not increasing, as not many new houses are built with thatched roofs nowadays, thatching still remains a healthy, thriving craft. It is estimated that there are still at least five or six hundred thatchers available in England. Young men are still being attracted to the craft, so the total numbers may even show a slight increase in the future. As there are approximately fifty thousand thatched dwellings in England, this means that each thatcher has an approximate potential average of one hundred thatched roofs to maintain and re-thatch. In addition, there are also hundreds of English inns which possess thatched roofs, and these can readily be seen in most parts of the country where the thatching trade prospered. There are also many thatched roofs still protecting farm buildings. Thatched roofs require much more regular maintenance than conventionally tiled roofs, so there is more than enough work for the existing thatchers. In addition, a few new houses are still occasionally built with thatched roofs, particularly in East Anglia and the West Country. Some of these have even won modern building design awards.

The demand for thatchers' services will no doubt last well into the future as there is a strong current desire to preserve, renovate and restore old properties. There is no reason why a thatched building which has stood for centuries should not last for at least another one or two, as long as there are thatchers available to keep the roof in good order. It is possible for young potential thatchers to start learning something of the trade by attending courses, sponsored through government retraining schemes. Also there exists the Council for Small Industries in Rural Areas (CoSIRA) which encourages the continuance of the thatching craft and has held courses in the art of thatching for many years. This council is always willing to offer advice and assistance. Its headquarters is based at Wimbledon in London, but many country regions possess area offices.

The usual practice for a prospective thatcher is to serve an apprenticeship and work for a number of years with an experienced master thatcher. Many counties have Thatchers' Associations, with their own appointed secretaries. Formerly, there was a society for these various associations of master thatchers, but the name has now been changed to the National Society of Master Thatchers, so that individual thatchers can be accepted as members in their own right. The society proposed in 1977 that a register of qualified thatchers should be compiled. It was suggested and hoped that only those who could perform to the high standards demanded of the profession and the craftsmanship associated with it, should be included and allowed to operate officially as thatchers. It is thought that such a register would protect not only the reputation of the master thatcher but also the general public against the increasing numbers of unqualified thatchers who now operate.

It is interesting to recall that it was precisely these two points which led to the institution of the Craft Guilds during the Middle Ages. At that time, a Craft Guild consisted solely of the skilled workers of an individual trade or craft, working together in close co-operation within a town or region. There were three types of members of a Craft Guild, masters, journeymen and apprentices. The length of service of an apprentice was often left to the whim of the master, but in 1563 a law was passed to specify a seven-year apprenticeship. At the end of this period, an apprentice was considered to be adept at his craft. He could then practise in his own right – but first as a journeyman. The name is derived from the French word *jour*, meaning day; a journeyman was thus paid for his work on a daily basis. He would often continue to work for his previous master until he could establish that he was an exceptional craftsman. This was done by the journeyman's making and submitting for approval a piece of his work which he considered to be of a very high standard, a '*master*piece'. On satisfactory acceptance of the work, the journeyman could then become a master of the guild.

The system became unacceptable when it was found that many highly skilled journeymen were unable, or could not afford, to become self-employed, their own masters. They had no alternative but to continue working for an employer. Due to this, the journeymen formed their own guilds, and by the seventeenth century these were established alongside the companies of masters.

Only skilled workers were members of the old craft guilds. Nowadays, an increasing number of unqualified workers undertake craft tasks in the thatching trade. Unqualified thatchers may offer apparently cheaper initial rates for general thatching work, but in the long term they will usually prove more expensive than the employment of a master thatcher. Even if the initial finished artistic appearance of the thatch is not taken into consideration, the passage of time may reveal faults in the roof. For example, heavy rain may cause sagging of certain areas, and this will entail costly repairs. Weak ridge construction will eventually result in a leaky roof. Poorly made gable ends will encourage bird damage. A good general guide to the ability of an individual thatcher can often be gained by asking the local inhabitants who live under a thatched roof, or by making enquiries with some of the farmers.

Thatchers in the various counties of England adopt their own individual distinctive finishes, particularly for roof ridges and also often for the gable ends. The thatchers of East Anglia, for example, normally have to make extremely high-pitched thatched roofs with steep, sharp gable ends, due to the architectural design of the houses in the region. Another distinguishing characteristic is that the bottom ridge edges are frequently finished with a great number of large scallops and points (Fig.48). In contrast, the thatchers of Devon often make roofs of a more chunky nature with a lower pitch, due to the particular roof design, and with more rounded gable ends.

Besides the major regional differences, individual thatchers

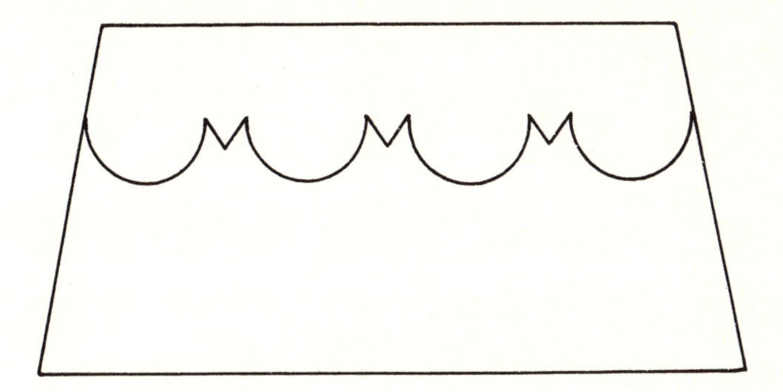

Fig.48 Scallops and points

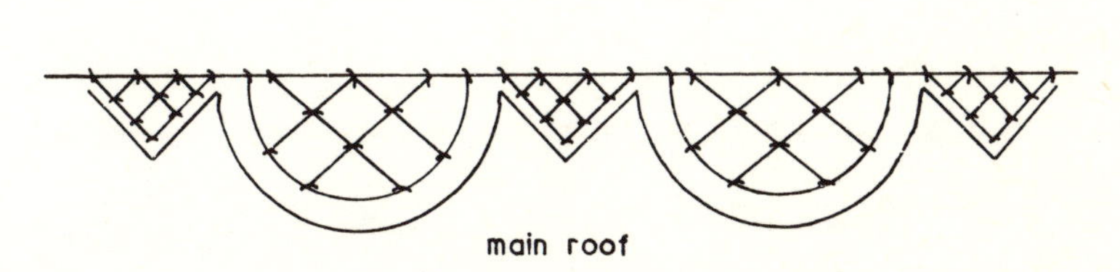

Fig.49 Decorated scallops and points

also make their own characteristic minor changes of detail. Instead of plain scallops and points, some may employ decorated versions (Fig.49). This involves the fixing of hazel lengths, spars and cross rods on the individual scallops and points surfaces to make patterns which are pleasing to the eye. Some may cut the occasional rectangle or square, rather than a point, along the traditional scallops and points design at the

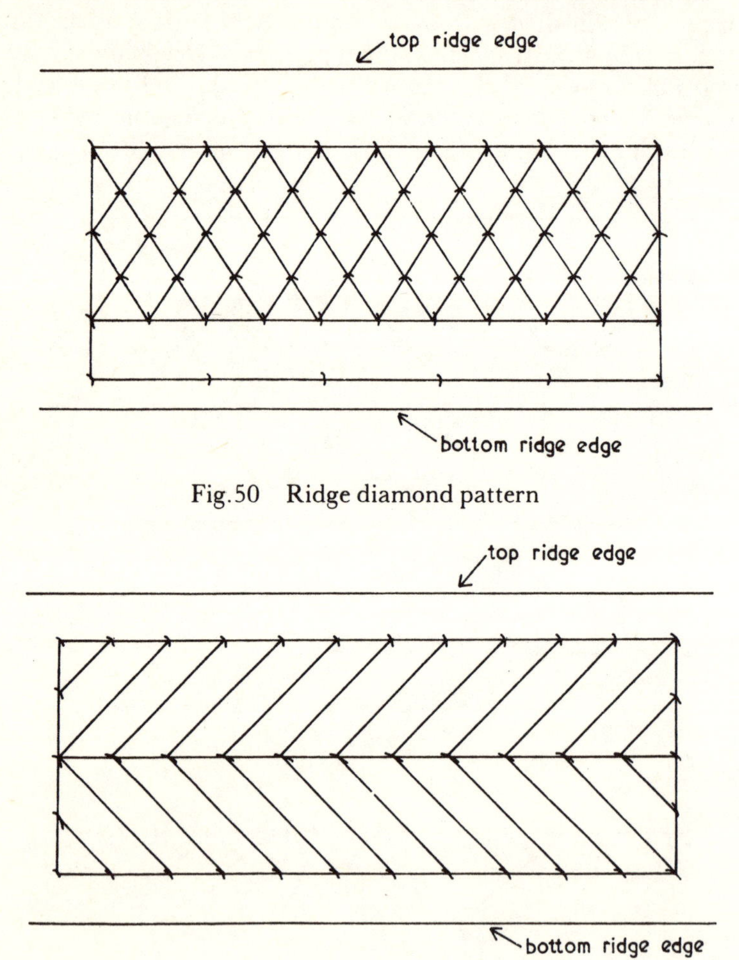

Fig.50 Ridge diamond pattern

Fig.51 Ridge herring-bone pattern

bottom ridge edge. In particular, rectangles are frequently utilized directly below the areas of chimneys. Others may cut a pattern in the bottom ridge edge consisting of alternating long and short points, in admixture with the occasional scallop. The cutting of long and short tongue shapes in the bottom edge of the ridge is sometimes favoured rather than the use of points. Along the complete central area of the ridge, some thatchers build elaborate diamond-shaped patterns, using horizontal hazel liggers and cross rods (Fig.50). There are many other variations, such as patterns made in the shape of dogs' teeth and also herring-bone (Fig.51). Some thatchers may use some form of ornamentation around the chimney apron area, in addition to that on the main roof ridge. At one time, the use of brambles interwoven around the spars was a popular decorative finish. Straw ropes were also often utilized to make various individual ornamentations to roof ridges and chimneys.

Thatchers also have their own characteristic methods for the trimming, shaping and finishing of the eaves and gable roof thatch edges. For example, they may be cut or dressed at different angles on the underface. The type of trimming employed around the area of the window spaces is also often the individual preference of the particular thatcher. In the

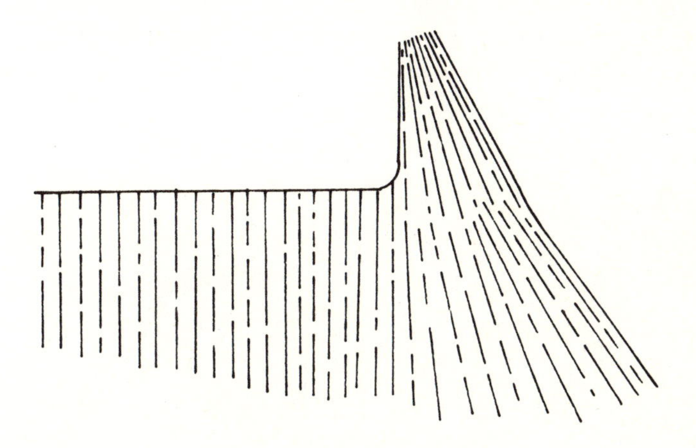

Fig.52 Large tuft pinnacle

case of long straw, the whole roof area may be severely trimmed down by certain thatchers to give an overall tidier and more distinctive appearance. Other thatchers will not adopt this procedure as they consider that a drastic trimming operation reduces the overall life of the roof, by the unnecessary removal of thatch material.

The type of thatch finish on the very top or pinnacle ends of the ridge on a house is also often a characteristic of the individual thatcher. Similar designs are often also used to finish the pinnacles of house porches. Some may be finished with a very pronounced large tuft (Fig.52). Other ridge pinnacles may be ornamented with a more gently and curved tuft (Fig.53). Some roofs may possess a long tuft with a zig-zag end (Fig.54). In Essex, the use of exceptionally sharp tufted ends is popular. There are a host of other variations in the different regions. It was once a common custom for some

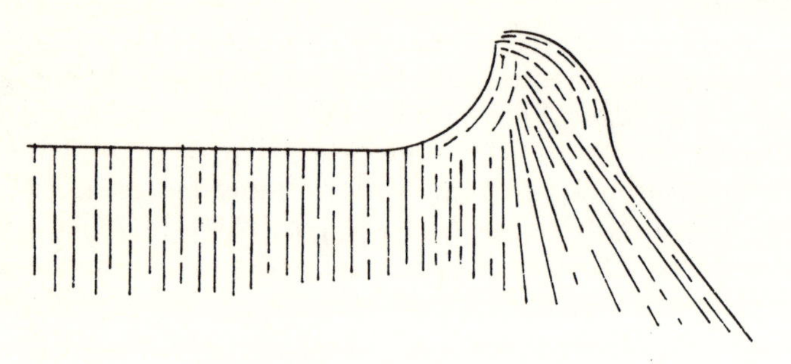

Fig.53 Curved tuft pinnacle

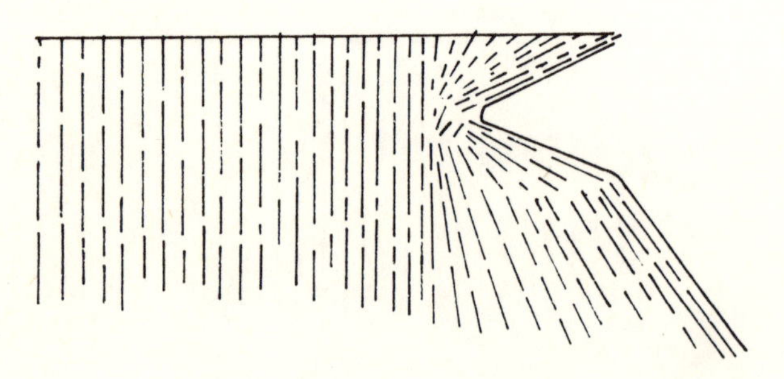

Fig.54 Zig-zag tuft pinnacle

thatchers to leave or hide a small tool or souvenir in the thatch layer, so that the next thatcher would discover it at some future date.

Distinctive corn dollies are also often placed on the tops of thatched buildings to decorate them. This custom is intended to bring good luck to the occupants, an extension of the original concept of ensuring good harvests and also the later use of dollies on hay ricks. However, corn dollies are also placed on buildings to show the pride of the thatcher in his craft and handiwork. They are often used to indicate which particular roof was thatched by a certain thatcher. They can therefore represent a thatcher's signature without the need for a written name. Most corn dollies used for this purpose represent animals or perched birds. Pheasants are a particularly popular variety seen on roof tops in the south and west of England. Peacocks are also sometimes seen, and foxes are popular in many country areas. Corn dollies representing fishes and also lambs are occasionally spotted. A dolly in the form of a farm tractor may be sighted, but this is fairly rare.

The straw used for making corn dollies, like that employed for thatching, must ideally be derived from the long straw wheat varieties. Again, the straw produced from the wheat grown nowadays for the combine harvester is not suitable. Many types have too much pith in their centres, rather than the desired hollow stems. Rye and oat straws are satisfactory if they can be obtained; barley straws are considered inferior. Skill is required in the making of a dolly, as the hollow straws have to be bent, twisted and shaped without causing them to fracture. If the straw has become dry in storage, it is essential for it to be wetted before use. This procedure can be avoided if the straw is used within about a week after it has been cut. Ideally, the wheat should be harvested when nearly ripe. In this state, the straw joint immediately below the grain ear will be green. The straw may be plaited, woven, folded and knotted during the construction of the dolly.

The technique used for the making of the particular type of dolly varies considerably. Some thatchers may encase a rough outline straw body in a wire mesh and then mould the outer wire case into the desired final shape. Others may adopt more elaborate procedures. These may involve the building of a basic skeleton shape in wire and hazel wood for a perched bird and then skilfully dressing the skeleton frame with tied and shaped straw sections. The final contours of the bird are obtained by trimming the straws into the various shapes required for the wings, tail and other body parts. Animals are normally built on a basic wire frame body. The rough shape of the straw animal is first constructed, and this is then encased in a wire mesh and finally shaped to the desired finished form. Rough dollies are sometimes quickly constructed by the thatcher using thatching straw, when standing on the top of his ladder, after finishing the ridge section of the roof. The dolly, which would perhaps look crude on a close inspection, will appear perfectly satisfactory when fixed to the roof top and viewed from the distance of the ground.

Thatchers have always been required to be versatile and to possess skills outside their own craft. On occasion, they have to turn their hand to basic structural repairs, so that they can replace, for example, a rotted roof timber with a new one. Additionally, it may sometimes be necessary for a thatcher to repair the brickwork at the bottom of a chimney stack before covering it with new thatch material. They are thus often called upon to use their own initiative and improvization skills as snags or unforeseen events occur during their routine work.

Thatchers are still sometimes called upon to thatch various items other than inhabited houses and cottages. These may range from the re-thatching of a huge preserved tithe barn roof, to the making of small thatched table umbrellas for garden use. These latter items are particularly popular in hotel and pub gardens. However, genuine thatched table umbrellas have a short life in these gardens, which are open to the public. This is due to children and many adults being unable to resist touching and pulling out straws from the bottoms of the umbrellas, which are always in the ready reach of extended hands. Synthetic thatched canopies are also frequently seen over bars in pubs and hotels. It is, of course, essential that the material utilized in these 'thatched bars' possesses fire-resistant properties, as many smokers congregate beneath them during licensing hours.

With regard to obtaining basic materials for their normal thatching trade, thatchers usually have to buy them from the grower. For example, a farmer may advertise that he is open to offers from thatchers for a certain acreage of special long wheat grown on his land. The price paid by the thatcher will depend on the market forces and the quality of the crop. The thatcher may occasionally undertake to harvest the crop himself. Some farmers grow and harvest themselves the long straw wheat varieties for the thatching market. The farmer then has to harvest it with an ordinary reaper and binder which entails high labour costs. This is primarily due to the additional handling required during drying, stacking and often later combing and tying the bundles suitable for the thatcher. The grain obtained from the wheat ears is kept by the farmer, who then offers the bundles of reed for sale. These factors mean that the thatcher has to pay a high price for his basic raw material. In the case of Norfolk water reed, the price will be even higher and perhaps additional expenses incurred due to the extra transportation costs, especially if the thatcher is a long distance away from the producing area.

Due to the high costs of present-day materials, thatchers have to be much more cost-conscious than their predecessors. Forty or fifty years ago, thatch was a relatively cheap material, and it could be used economically for the roofing of most objects, such as cottages, barns and haystacks. Today the thatcher is dealing with a much more expensive commodity, and the market for his craft is approaching the near luxury category. The time required for thatching a roof, the high insurance rates and the initial cost of the thatch material all tend to make a thatched roof an expensive proposition to a potential customer. It is therefore extremely important for a thatcher to be able to estimate the exact quantity of material required for a particular roofing job before quoting a firm price to a customer. An accurate measurement of the roof area to be thatched has therefore to be made by the thatcher. Some thatchers will quote a price for thatching without supplying the thatch material. This may be preferred by a farmer, for example, who is fortunate enough to be able to supply his own material for roof thatching.

Other material costs have to be estimated by the thatcher, such as the number of thatching spars, liggers, sways, etc which will be needed. The cost of a protective wire netting over the thatched roof may have to be included in the basic material costs. There are also the transport costs, general equipment and miscellaneous overhead expenses of the thatcher's business to be accurately assessed and the cost to be charged for his own labour and any apprentice employed. The estimated time to complete a thatching operation will vary depending on the architectural design of the house. It will be influenced by such factors as the number of valleys and dormer windows present. These will require additional thatching time to normal straight roof section thatching. The accurate assessment of the total time spent, from the arrival to the final departure from the site, is essential in order that the thatcher does not undercharge for his labour. Thatchers may have to submit a separate quotation to a customer who requires an ornamental raised ridge. This involves the use of more thatching material and more labour costs.

Labour charges vary, and the more skilful and better the reputation of the thatcher, the more he is entitled to charge. From the customer's point of view, it is worth recalling that the best thatchers are normally the cheapest in the long term. Although the initial price may be more expensive, the later maintenance and repair costs will probably be much reduced. The reputations of the various thatchers in the area have usually been well established with the local customers over the years. Fortunately, from the customer's point of view, there will normally be more than one thatcher in a particular area, although the one chosen may not always be available at short notice.

It is difficult to put an exact price on craftsmanship. However, thatchers, like all other workers, must get an adequate return for their efforts. The total business capital investment involved is not great with a thatcher, so it is not possible to calculate a reasonable return on this particular aspect. An adequate return must therefore be based on the hours worked and an additional reward obtained for the thatcher's experience and the artistic nature and craftsmanship of the work involved. Much of this craftsmanship depends on the individual artistic judgement of the thatcher's eye and hand. Few scientific measurements and instruments are used by the thatcher to assist him in his work. No exact lines are predetermined. A good thatcher knows instinctively what is correct to obtain a good practical waterproof roof and at the same time one which has a form pleasing to the eye.

5
Living in a Thatched Building

Most people who live in thatched properties take great pride in the appearance of the thatch and its aesthetic appeal. Also, most lavish much attention on maintaining their properties in immaculate order. This is one reason that thatched houses generally command a good market price and can be sold fairly readily when required. The market may be more specialized than with more conventional houses, but estate agents, in normal times, encounter no difficulty in finding prospective purchasers who are looking for thatched properties. At one time, many building societies were very reluctant to lend money to people who wanted to live under a reed or straw roof. This prejudice against such buildings has now been partially overcome. In fact, it has always been possible to raise a mortgage with one of the smaller building societies.

The previous marked reluctance to lend money on such properties was based not only on the presence of a thatched roof, which would require more maintenance and expense than a tiled roof to keep in good order, but also on the fact that most thatched properties are old. Walls, for example, may have been made of such materials as cob or chalk ashlar. It is known that there is a particular risk with these materials, if at one time there had been a leaky roof, which allowed water to get into the tops of the walls. The water may have frozen during a cold spell in winter, and the walls would then have suffered deterioration. However, as mentioned earlier, cob and chalk will last for centuries if kept dry. It was also likely that the beams and timbers in the house may have been subjected, over a long period of time, to rot conditions because of the absence of a damp-proof course and a roof in disrepair.

All these factors contributed to the fact that the property may not have formed a sound enough security for the money to be loaned, especially if a compulsory resale was ever needed to discharge the mortgage debt.

The picture has been considerably altered in present times. Many more building societies will now lend money on older thatched properties, if the borrower agrees to undertake any stipulated repairs or improvements. For example, these may be the improvement of the damp-proofing of the house and repairing or renewing the thatched roof. The building society may withhold a portion of the mortgage money until suggested alterations have been completed to their satisfaction. The house-owner will also have to undertake to maintain the thatched roof in good condition, which will mean that he will be involved in some additional future expenditure besides having to meet his mortgage commitments.

It is always possible for a prospective owner to arrange for a local thatcher to survey the state of a roof on a property before making a final decision to purchase. The local council will also probably be able to help arrange a roof survey on the payment of a standard fee. The condition of the thatch and the likely future expenditure on the roof, depending on its particular state, will then be known. Long-term overall economy will probably be best achieved by using the most skilled thatcher available in the area to carry out the survey and also to employ him for any future work, if the house sale goes through.

The prospective purchaser's solicitor will determine during his searches whether the thatched property is listed officially as a building scheduled as being of historic or architectural importance. This has both advantages and disadvantages. For example, it may sometimes be possible for financial assistance to be obtained towards the maintenance of the building, from a government or local authority grant. On the other hand, considerable delays must be expected in obtaining permission to make any desired alteration to the exterior of the building. It may even be deemed that external changes are not permissible.

There are several advantages to living in a property with a thatched roof, in addition to the pleasing appearance and the irresistible charm it holds for a great number of people. Most thatched dwellings are old and possess relatively large gardens. They were mostly built on the best possible locations which, in past days, could be selected from a wide area. The original sites were probably chosen because of their nearness to a natural water supply, together with their aspect and the possible shelter from the weather which could be obtained from the surrounding land contours. The buildings would usually have been built on slightly sloping ground to avoid or reduce damp problems. This former wide choice of sites contrasts with the present-day availability of building plots, which are often not in ideal locations and possess only small gardens.

However, there are also sometimes disadvantages with the siting of old properties. They may have been built for convenience near a track or road because the former occupants would not have possessed their own transport. These tracks may now have developed into a main country road on which busy holiday traffic may pass. A few thatched cottages may be situated in very remote country areas, away from villages, and they may not always be connected to a supply of mains tap water. The sole water supply to many such homes is frequently taken from private water wells, often situated in the gardens of the individual houses. Alternatively, several houses may sometimes share one well. However, the problem is not too serious, as the water can be adequately purified by the installation of a small chlorination unit or a suitable filtration system.

Another disadvantage of a remote property may be that it is not connected to a mains sewage drainage system. A cess-pit or septic tank may be installed. Septic tanks are less trouble than cess-pits and are therefore preferable. Cess-pits will require frequent emptying. Septic tanks, on the other hand, will require little attention over several years, provided that they are situated well below the house level and are buried in good draining soil. It is also prudent not to allow strong doses of disinfectants or detergents to go down the sink, in case they affect adversely the essential bacterial colonies which make the septic tank function efficiently. A further disadvantage of a remote rural property, especially when first lived in by a former city- or town-dweller, will be the apparent intense darkness outside the house because of the absence of street lighting at night. This can be partially remedied by the installation of outside lights, especially near the garage parking area. These disadvantages of living in a remote property apply, of course, to all types of houses whether they are thatched or not.

However, there are many advantages to be gained by living in a thatched property. The thick layer of reed or straw on a thatched building possesses an excellent sound-deadening effect, which is a decided advantage in these modern days of overhead aircraft noise. The clatter of vehicles and farm machinery is also effectively masked. In addition, the thick thatch layer has a good heat-insulant ability, which results in a house remaining warm in winter and cool in summer. The coolness of a thatched house on a very warm summer's day is particularly noticeable and refreshing. The insulation ability of a reed or straw thatch is due primarily to the very large numbers of hollow tubes, filled with air, which make up the thickness of the layer.

As a high proportion of the heat in a house is normally lost through the roof, there will be a worthwhile saving in fuel costs during winter. There is also no urgent need to insulate the loft floor with a three-inch-thick blanket of fibre glass or rock wool, as is normally a necessity in the conventional tiled roof house, because, without insulation, about twenty-five per cent of the total house heat is estimated to be lost through a tiled roof. The space in the loft of a thatched house remains at a reasonably constant temperature level, and the danger of any water pipes or storage tank situated there freezing in winter is much reduced. Of course, it is still advisable to lag any pipes in the loft area as an additional safeguard against the frost in a really severe winter.

Much heat is lost in the winter through the windows of houses. In modern houses, with large windows, the heat loss may be as high as twenty-five per cent of the total house heat. Generally, thatched properties are old, and they were originally built with much smaller window spaces than modern houses. A further saving, with regard to heat losses, is thereby achieved due to the smaller total surface area of window panes exposed to the outside atmosphere.

Most old thatched houses have solid or thick rubble filled walls, frequently constructed of stone, chalk or cob. They therefore do not have an exterior wall separated from an interior wall by an air gap. This type of cavity wall is a relatively modern building technique, designed to prevent the inner wall becoming damp when the exterior one is wet. In practice, the two walls are not entirely isolated from each other. Metal wall ties are usually between them as a support, but these are designed not to conduct water from one wall to the other. However, sometimes, during the construction of the brickwork, a little mortar is accidentally dropped into the cavity between the two walls, and this can adhere to the metal ties. The small pieces of mortar can then act as bridges and allow a small amount of water conduction between the walls. This can result in a few damp spots on the interior wall.

With regard to relative heat losses, the greater thickness of the solid wall type of construction, provided it is dry, will normally have the advantage over the cavity wall. The rate of heat loss through a solid wall will depend on the relative

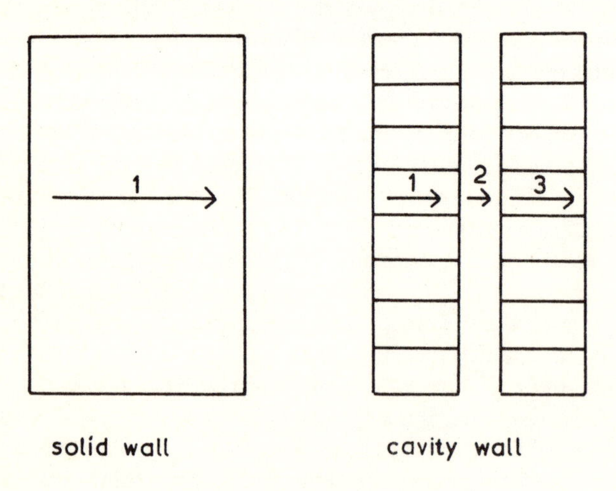

Fig.55 Wall heat losses

temperatures of the room and the outside air, or in other words the temperature gradient across the wall. The thermal conductivity and the thickness of the wall will also affect the heat transfer rate. The heat is lost through the wall by conduction in one stage (Fig.55). With a cavity wall, the heat is lost in three stages. The small effect of the metal ties and mortar bridges which may be present between the walls can be ignored. Firstly, heat is transferred through the inner wall to the air gap, then through the air gap itself and finally through the outer wall to the atmosphere. However, as the combined thicknesses of the interior and exterior walls are normally much less than that of a solid wall construction, the heat losses will, as mentioned, normally be greater with a cavity wall.

For this reason, it has become a modern practice to insulate cavity walls to save heat losses from the home. In the process, an inert heat insulant urea-formaldehyde foam is pumped into the cavity. This displaces the air in the gap, and the insulation layer is left in its place. Great care has to be taken in the cavity-filling operation to ensure that all the air is displaced, so that the entire space is filled. Alternatively, cavity walls are sometimes filled with mineral wool. It has been estimated that cavity wall insulation can help to save twenty per cent of the heat loss from a house.

In the summer, the reverse applies with regard to the relative heat transfer rates. The thick solid walls usually found with thatched properties, and the thick layer of thatch material, slow down the rate of heat flow into the house from the hot air and sun outside. The house therefore remains cool. The smaller window space area also assists in the process. It will be particularly noticeable that any water stored in a cold water tank in the loft under a thatched roof remains cool during hot summer days. With a tiled roof, in comparison, the stored water quickly becomes warm.

In contrast to modern houses, many old solid wall thatched cottages will not possess damp-proof courses or dug foundations. It is probable that the interior of the walls may become damp by moisture rising slowly up the walls from the ground. The use of a damp-proof course in modern houses

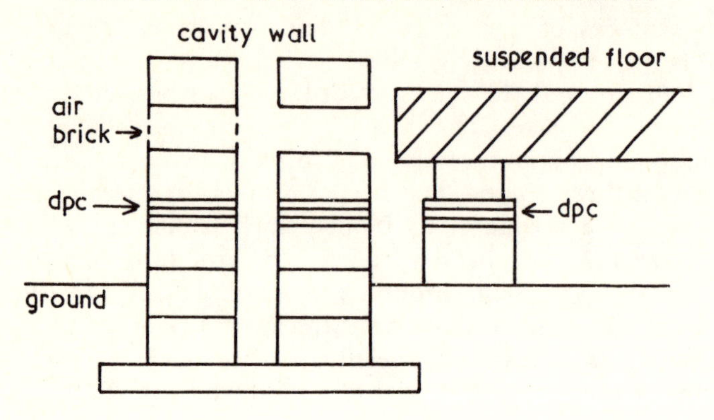

Fig.56 Damp-proof course

prevents this occurring. A damp-proof course is normally a water-impervious layer of material placed into the brickwork course, approximately six inches above ground level around the entire building (Fig.56). Unless the damp-proof course is short-circuited in some way, it is not possible for the dampness to rise above it in the walls, and the house is kept dry.

Modern houses also often possess suspended ground floors, and because of this arrangement air can circulate underneath through air bricks. Old cottages are frequently of solid floor construction and are built directly on the earth without a damp-proof course. Water therefore can rise slowly through the floor as well as the walls.

Despite the lack of damp-proof courses, the development of modern building materials and techniques allows vast improvements to be made with regard to reducing damp in solid wall and floor dwellings. This is possible even if a building, at first glance, appears to be in a derelict condition. Many different methods are available. For example, it is possible to have a completely new ground floor installed. This can be done by first laying hardcore, after the excavation and removal of the existing ground floor material. The excavation process also gives a useful opportunity to obtain some additional head room. (Old cottages are noted for their low

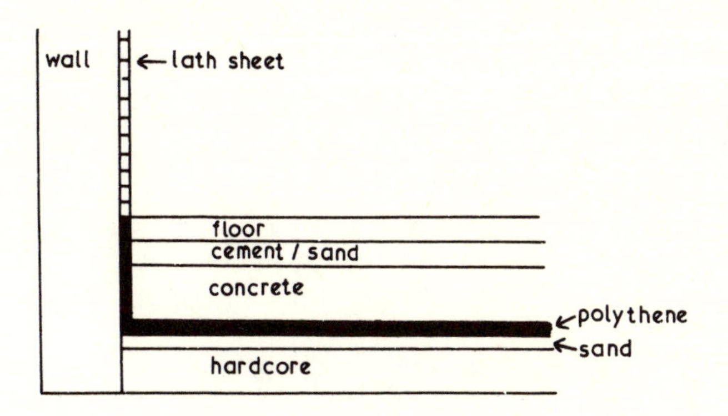

Fig.57 Solid floor and wall

ceilings.) The hardcore is then covered with sand, in order that a membrane layer of a tough thick-gauge polythene sheet can be bedded down on it (Fig.57). This waterproof membrane, which acts as a damp-proof course, is then covered with concrete and finished with a cement and sand screed to receive the new floor finish. Provision is, of course, made for the laying of service pipes and drains in the concrete layer. The edges of the polythene membrane are turned up to form a waterproof joint with special damp-proof corrugated lath sheets which are fixed on the surfaces of the interior walls. These sheets are then plastered over to make a new interior wall surface. This flat new surface may not be considered as attractive as the quaint unevenness of the original walls, and the charm of these will be lost to some extent.

The dryness of the house can also be improved by the excavating of soil away from the exterior walls of the dwelling. Soil has a tendency to build up slowly, over the years, around the outside walls. The laying of a concrete apron around the dwelling to make a fall-away channel to carry rain water dripping from the thatched roof to gullies, leading to soakaways, further improves the situation. Soakaways have to be used when the property is not connected to a main sewage drainage system.

With regard to dryness, a thatched roof maintained in good

condition provides a very good weatherproof cover to a house. It can withstand gale-force winds when they are not sustained over a long period of time. It rarely leaks, and, unlike conventional roofs, there are no tiles or slates which can slip to let in water. The occasions when leaks do occur with thatched roofs are normally rare. For example, it sometimes happens when one of the wooden spars, used in the ridge, becomes broken and turns into the thatch and penetrates it. The broken spar can then create a channel to allow water penetration through the thatch layer.

Old thatched roofs sometimes develop leaks after a prolonged spell of dry weather. This is because old compacted thatch patches have a tendency to open up or breathe slightly under very dry conditions. When rain falls again, this may lead to the occasional trickle of a few water drops through the thatch. However, this fault is normally self-healing, as the compacted thatch has a tendency to close up again when it becomes damp. Another potential leak-point for a thatched roof is at its junction with the chimney, where the thatch is likely to compact away from the brick or stone work. Birds may also have helped to pull some of the thatch material away in this area. As a temporary measure to prevent further damage, it is advisable to fix some wire netting over the affected area. However, if a leak occurs, it is best dealt with immediately by a thatcher. He will be able to overlap the chimney flashing with new thatch material.

When rain falls, thatched roofs are designed to shed water fast, so that the dwellings they cover are kept very dry. Yet, perhaps one of the greatest disadvantages of a thatched roof is, in fact, when rain falls. Water cascades down the roof and drips continuously, around the whole perimeter of the eaves for a considerable time, even after the storm has passed. A caller can become drenched, waiting patiently outside the house for a door to be opened, unless a porch is provided as a shelter. Without a porch, the caller might have to stand under the edge of an umbrella or tree, with the water drips constantly falling on his head. Although the eaves of a thatched roof have considerable overhang from the walls, they do not, unfortunately, provide shelter at ground level. Porches can be built in an attractive manner, and they are best thatched themselves to blend with the appearance of the main roof.

The water-dripping phenomenon with thatched roofs is accentuated because rain-water gutters and downpipes are rarely employed. There are two main reasons for this. Firstly, the steep roofs are designed to allow very fast water-shedding rates, which would rapidly overflow standard collection-gutters during heavy rainfall. Also, gutters are not really needed, as the thatched roof is given a good overhang at the eaves, to throw the water well clear of the walls. Secondly, thatch material was traditionally used long before man developed cast-iron and plastic gutters. Also, their possible present-day use may considerably detract from the appearance of the thatched roof. On the rare occasions on which gutters are utilized, they are of a much wider construction, due to the large overhang, than the standard rain-water collection-gutters seen on tiled roofs. The special gutters are often made of wood because of the large width. The normal absence of downpipes and gutters has one advantage in that it obviates the chore of cleaning, maintaining and painting them. A disadvantage of the absence of gutters is that the rain water, when it cascades from the eaves of the thatch, causes much splashing of the bottom walls around the whole perimeter of the house. If the walls are white, then the resulting mud spots will soon detract from the appearance. Some people attempt to overcome the problem by painting the bottoms of the exterior walls black. However, constant water splashes from the ground to the bottoms of the walls can encourage the growth of patches of green mould. This growth can be fairly easily destroyed by the occasional washing of the bottom walls with a commercial bleaching agent diluted with water.

The open-tube structure of reeds and straws ensures that they are not subject to self-ignition, as can sometimes occur, for example, with hay. However, thatch will burn if ignited by an outside source, and it will be found that insurance companies will consider the fire risk with thatch to be considerably greater than with other roofing materials. They will thus inevitably charge relatively high fire insurance premiums on thatched roof properties and also for the household contents. Some thatchers express doubts whether there is a greater fire risk, especially when a modern chimney design is used in a house. Old-time thatchers were sometimes known to pour a bucket of red-hot coals or roll a burning log over the top of a newly thatched roof, to demonstrate that there was no exceptional fire hazard. The available statistics suggest that the percentage of fires with thatched roofs is no greater than with other roof types. Most thatched cottages have stood for centuries, and more are pulled down by demolition than burned down. Many thatch fires are caused by sparks from extraneous sources, but it is unusual for the sparks from the house chimney to cause a roof fire. More likely are fires which are started inside the roof space, due to a faulty or grossly overheated chimney stack. In the event of a fire breaking out, it is likely that a greater total amount of damage will occur, as straw and water reeds are combustible products, while tiles and slates are not. This means that the potential liability of the insurance company for damage would be greater, as the fire would be likely to spread more rapidly. This is especially so when the thatch material is dry and air can penetrate below the surface layer through bird holes in the thatch. It is worthwhile to 'shop around' with several

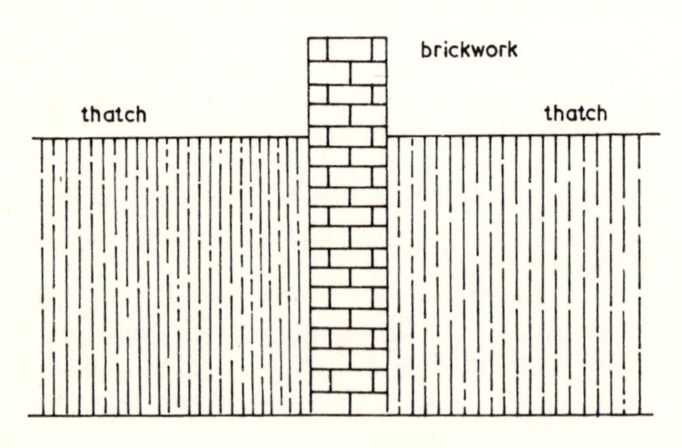

Fig.58 Roof separation

insurance companies to obtain premium estimates before making a commitment to a particular company. An alternative way is to use a reputable insurance broker, who will do this preliminary ground-work. He will be more likely to know which insurance companies look most favourably on thatched properties.

The amount of fire insurance premium demanded will depend on the type of construction of the thatched property and also on whether it is isolated or attached to a neighbouring house. If it is attached to another house, it is normally considered preferable if there is a raised brickwork ledge separating the two roofs (Fig.58). A further consideration may be whether the property is connected to a supply of mains water, so that an ample amount will be available for fire-fighting purposes, should the need ever arise. Mains water is always preferred to other sources, as it is supplied with a reasonable immediate pressure. It takes valuable time to arrange equipment to pump well water, for example, with an adequate pressure for fire-fighting purposes. Also, the fire brigade are limited in the quantities of water they can carry by wagon to fight a fire. The distance of the thatched property from the nearest fire station and the ease of access from the road will also be factors noted by the insurers. Some roofs are completely covered with wire netting for protection against bird damage, but this may lead to higher fire insurance rates, because of the added difficulty in pulling the thatch material away in the event of a fire-fighting operation. The wire netting hinders the procedure because the thatch material is underneath it. Time will be lost in removing the wire, despite the fact that the individual sections are normally fitted together so that they can be easily released in an emergency.

The presence and use of an open fire in a thatched property, will also have an appreciable bearing on the fire insurance premium. Another factor will be whether the chimney is new or old and whether it is regularly swept. The position of the chimney will also have an influence on the potential fire risk. It is best if the chimney is tall and is positioned at the highest point of the thatched roof. Local council regulations and insurance companies often require chimneys to be at least four feet tall and preferably six feet above the roof level. Open fires in the house are then considered safer, as any sparks emitted from the chimney have time to become extinguished, before they can fall on the thatched roof itself. It is also considered safer, as a precaution against fire, if the chimney stacks are constructed with walls at least nine inches thick.

The fixing of a manufactured spark-arrester to the chimney of an open fire is not always an easy practicable proposition. Spark-arresters are frequently encountered in industry as fire safety devices when fitted, for example, to chemical plant chimneys, funnels and industrial engine exhausts. They contain a corrosion- and heat-resistant fine wire mesh and are designed to stop any large solid sparks from escaping into the atmosphere. Such devices are not commercially available for fitting to private house chimneys. However, it is possible to find the occasional company prepared to make a spark-arrester to an individual's own specification. This will be manufactured to suit the particular interior diameter of the chimney-pot and, when fitted to the chimney top, the spark-arrester will look like an inverted wire-mesh bucket (Fig.59).

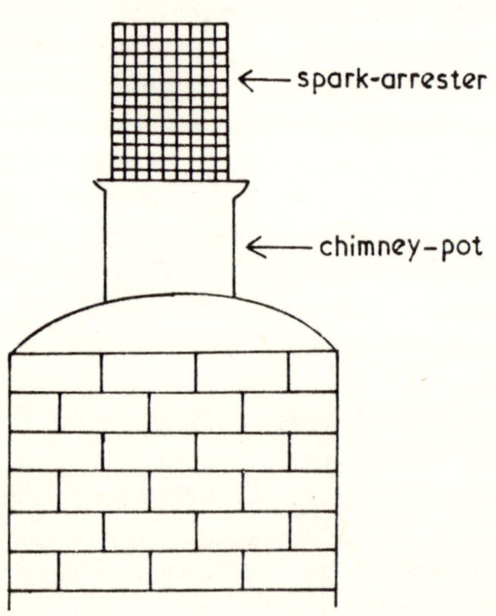

Fig.59 Spark-arrester

Legs attached to the spark-arrester fit into the top of the chimney pot. The Fire Protection Association in London may be able to assist in indicating a suitable firm prepared to manufacture such a device. However, difficulty may sometimes be encountered in obtaining a suitable type, which will not cause a smoky fire due to the gradual clogging of the mesh. An ordinary fine-wire mesh fitted directly over the chimney would quickly become clogged with soot and create a smoke problem in the house.

It is always advisable not to use paper to light or 'draw' fires in an open hearth but to use instead fire-lighters and wood tinder. Paper can easily be sucked up the chimney with the draught and fall, still alight, from the chimney top onto the thatch. This type of accident is avoided if a spark-arrester has been fitted to the chimney. It is very advisable to have chimneys regularly swept in order to reduce the risk of chimney fires. The burning of coal produces a different type of soot from logs. The coal yields a softer, fluffier type, which can be brushed off the surfaces of the inside of the chimney relatively easily with a sweep's brush. Wood logs produce a much more adherent, harder type of soot, which normally has to be removed from the inside of the chimney by the sweep using a scraping tool rather than a brush. Certain wood soots can be much more inflammable than coal soot. However, the total amount of soot deposited in the chimney by the burning of logs is much less than with coal, and the chimney will not have to be swept quite so frequently. Logs are therefore perhaps slightly safer to burn than coal in a thatched property, provided that the adherent soot coating is scraped off fairly regularly from the chimney surfaces. The best logs to use to get a cheerful and hot fire are ash and oak. The latter are much improved by storage and weathering. Some people burn a mixture of coal and logs. The coal, in particular, is used to create a good initial heat before the logs are put on the top. Most thatched properties are situated in country areas where there are no smokeless zone regulations. The owners of the few which are sited in the near suburbs of cities have, of course, to abide by the local regulations with regard to the type of fuel which can be burned.

Many old thatched cottages have chimneys which were deliberately constructed to be non-vertical. They will be seen leaning slightly inwards, and the alignment is incorrect. This was done to ensure that in the event of the chimney accidentally falling, it would topple inwards and land on the roof, rather than outwards where it would fall on the ground or, worse still, on the head of a passer-by. It will also be discovered that most old chimneys have not been lined or sealed efficiently with a chemical fire- and smoke-impervious layer throughout their entire lengths. This means that when an open fire is burning, there is a possibility that traces of smoke fumes may escape into the upper rooms of the house. This is especially so if the brick or stonework of the chimney has become badly cracked with heat and age. Also, the acidic combustion gases from the fire can penetrate and spoil, by staining, the plaster work and the wallpaper in the areas surrounding the chimney breast. This is particularly the case if the chimney is damp, due to rain entering the porous stack. These faults can be completely overcome by having the chimney lined, but this is normally an expensive process, due to the high cost of the lining material which has to be both chemically inert and fire resistant.

While on the subject of chimneys, it is worthwhile to

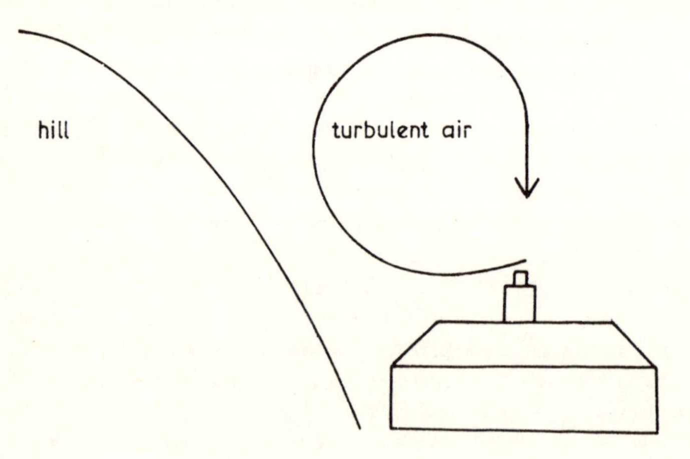

Fig.60 Chimney down draught

remember that if there are any exceptionally high buildings, trees or hills in the immediate vicinity above the thatched property, then there is a possibility that an open fire may have a tendency to 'smoke' more than usual. The high objects may cause a down draught in the chimney, through the creation of turbulent air, and some of the smoke, filling the chimney, can be pushed down into the room (Fig.60). The wind direction will have an important bearing on the matter, but it is possible to alleviate the smoke problem, to some extent, by experimentation with a suitable chimney cowl.

Further on the subject of thatched roofs and fire risk, it is well to be cautious when lighting bonfires. Before incinerators for burning rubbish are lit in the garden, care should be taken to ensure that the wind is blowing away from the house. It is also preferable to light bonfires early in the morning, when the roof is still covered with dew. If this is not convenient, it is best to wait until it has rained and the thatched roof is wet. Although the bonfire itself will also be damp, it can still be ignited with the use of diesel oil, paraffin or solid wax fire-lighters. Fireworks should, of course, never be lit in the vicinity of a thatched property. Also, great care must be taken if a blow-lamp has to be used for plumbing purposes. This is exceptionally dangerous in the loft area. Blow lamps should never be utilized for burning off old paint from windows or doors of thatched properties.

It is now possible to obtain special chemical solutions which can be used to treat thatch material to give it improved fire-resistant properties. The solutions contain fire-retardant salts which cut down the oxidation rate of the thatch material, if it should catch fire. This has the effect of slowing down the rate of spread of the fire. However, the thatch material must be thoroughly impregnated with the chemical salts, otherwise their action will not be of a long-lasting nature. The solutions have to be applied to the straw or water reed before the roof is constructed and must be allowed ample time to soak into the thatch material and to dry thoroughly. The fire-retardant solution cannot be applied after roof construction, because the overlapping layers of thatch material would prevent its reaching the underlying reeds or straw. Due to the thatched roof rapidly shedding water, the solution would quickly run off, and no penetration into the thatch material could occur. A disadvantage of the use of fire-retardant chemical solutions on thatch is that they sometimes encourage the growth of moss and moulds.

If the timbers in the loft which support the thatched roof are robust enough, it is also possible to screw or nail sheets of a fire-resistant material, such as plaster board to them. In the event of a fire, this procedure would assist in keeping the fire outside the house. Another beneficial device, to assist possible fire fighting, is the fitting of a long perforated water pipe along the ridge of the thatched roof. This can be employed only if the thatched dwelling has a mains water supply to which the pipe can be connected. Sufficient water pressure is essential to reach the top of the roof, preferably from an outside mains water source and stop-cock. The sprinkler device is also a preventative measure as it can be used if there is an imminent risk of a fire. It is even possible to obtain completely synthetic fire-retardant thatch material, but it looks false and lacks aesthetic appeal. At the moment it has not significantly replaced the traditional naturally grown water reeds and straw.

As mentioned earlier, thatched roofs are prone to attack by birds, especially when they are searching for nesting places or materials. They also do damage when searching beneath the eaves for insects or for seeds left in the wheat straw. (Sparrows, in particular, are a nuisance. They were sometimes called 'thatch birds' by old country folk.) The attack results in the formation of holes in the thatch, which look as if they were actually punched into the thatched roof. Greater repair costs are inevitably encountered than with conventionally tiled or slated roofs which, of course, do not suffer from this disadvantage. To keep the costs to a minimum, it is prudent to repair holes as soon as they are noticed, by pushing handfuls of straw into them. If this is not done, the birds will tend to concentrate on the holed areas and enlarge them still further. This temporary repair should be done only if the holes are fairly accessible and can be reached without damaging the thatched roof. If it is decided to risk the

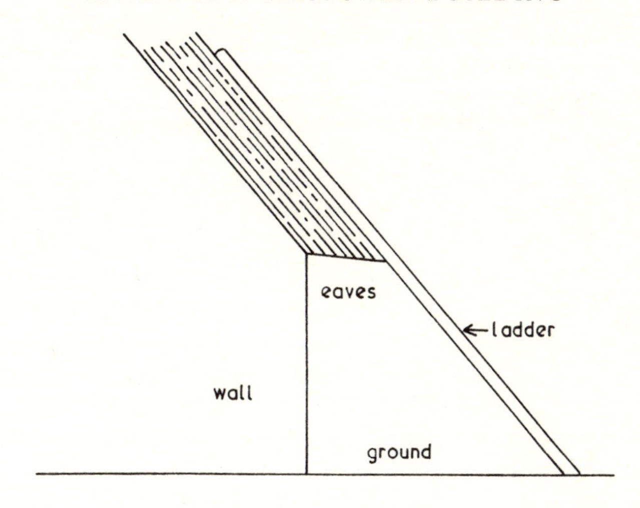

Fig.61 Ladder on roof

use of a ladder to reach the holes, it is essential that it be positioned accurately, so that it lies absolutely flat on the surface of the thatched roof to spread the load (Fig.61). If the ladder is placed at an angle to the thatched roof, it will cause severe damage by breaking down the eaves. It is also advisable to push in any isolated individual reeds or straws which may have been only partially pulled out of the thatched roof by the birds. These can normally be easily reached, where they are mostly found, along the eaves and around the window spaces. However, it may not always be possible to push the individual straws or reeds back into position, because of the tightness of the thatch layer.

Birds enjoy seclusion and shelter and usually attack the thatch at places where they can hide from view. There are some bird-scarers available, but their usefulness is doubtful. Also, many people like to encourage birds into their gardens, and so much depends on individual feelings on the matter. Favourite spots for bird damage are under the eaves and gables and at junctions of the roof and chimney, where the thatch material has a tendency sometimes to compact slightly. Some roofs are wire netted at these strategic points to prevent bird damage. Many roofs, especially those constructed of long straw, are completely covered with wire netting for protection, but this may lead, as mentioned, to higher fire insurance rates because of the added difficulty of pulling the wire and thatch away in the event of a fire. The wiring of a thatched roof, of course, costs money, and this is another factor to consider.

The growth of trees and bushes near a thatched roof must be carefully watched. Pruning must be carried out at regular intervals to prevent branches rubbing against the thatch material and causing rapid deterioration by abrasive mechanical damage. The close proximity of trees may also give rise to leaves and twigs falling on the roof, and if this is the case, it is not advisable to have the roof covered with wire netting. Trees also cast shadows and cause additional water to drip on the roof.

These damp conditions can sometimes favour the development of moss. However, moss formation will depend on the prevailing wind direction to which a particular roof surface is exposed. It will also be affected by the nearness of hedgerows, together with the amount of sunshine which can illuminate the roof. Mosses have no true roots and are attached to a roof by hairlike threads called *rhizoids*. Moss spores reach the roof area, when spore dispersal occurs during dry conditions, and the wind direction is therefore of importance. Large areas of well-established moss on a thatched roof can greatly interfere with the water-shedding properties and encourage the thatch material underneath to rot. It also spoils the appearance of a thatched roof. The only dubious advantage of a thick moss layer on a roof is that it may make it slightly more fire-resistant. On some very old thatched roofs which have been allowed grossly to deteriorate and rot, grass may even be seen sometimes growing on small areas of the roof surface, as well as the moss.

If ivy is allowed to grow up the wall of a thatched building, it is always advisable to cut it below the eaves level. This stops the ivy spreading to the thatch material itself and so prevents the accelerated deterioration of the roof and interference with its water-shedding properties. It is advisable with old properties, especially those without damp-proof courses, to cut the ivy only below the eaves level and not to remove it completely from the walls by severing the base of the plant. This is particularly so if the ivy is well established over the whole wall area. Ivy fixes itself by aerial roots, and the moisture the plant is able to absorb can assist in keeping walls dry. The mass of ivy leaves also prevents a lot of rain water from directly hitting the wall. These processes may prevent moisture penetrating to the interior wall surface of the house.

A further disadvantage of a thatched roof is that it has a tendency to release small pieces of straw and dust, which descend into the loft area of the house. This is a particular nuisance when a water-storage tank is situated in the loft, and great care should be taken to ensure that it is adequately covered. The presence of the thatch material in the roof also attracts and encourages spiders to make their homes in it. This means that many cobwebs will be formed in the loft area, and these will be also noticeable around the outside window areas of the house and under the eaves. Fortunately, all the spiders in England are of the non-poisonous variety.

Loft spaces were originally rarely built in thatched houses, and such spaces which have been later incorporated are usually very cramped with little headroom. They are also very poorly ventilated, and this should be remembered when materials such as woodworm killers are applied to roof timbers. These products are made up in kerosines or volatile low-viscosity solvents, so that good penetration of the chemical into the wood can be achieved. However, in a restricted space, vapour may accumulate which can be dangerous on inhalation. The hydrocarbon mist can also present a fire risk. Such work as woodworm treatment is therefore best done in small stages to allow the vapour ample time to disperse. Light switches should not be used for a considerable time to avoid sparks causing a possible fire.

In order to ensure that the walls of a thatched dwelling remain dry, when it is raining and water is being shed from the roof, it has been seen that the eaves always overhang the outside walls by a wide margin. This design feature unfortunately has a tendency to restrict slightly the amount of light in the upper rooms of the dwelling. It is therefore best to use light colours for interior decoration in upstairs rooms in order to optimize the amount of light available. It will also often be found that the rooms of a thatched house are fairly narrow. This is because the steep roof pitch required can be obtained only if a relatively narrow span is used. As mentioned earlier, the building of long thatched houses overcame the room space problem to some extent.

It will appear that the disadvantages of living in a house with a thatched roof outweigh the number of advantages. However, the disadvantages are relatively minor, and the advantages are considered by many to outweigh them, especially by those who prefer charm, quaintness and oldness. The warmth and aesthetic appeal of a thatched roof are of inestimable value to many owners and prospective purchasers. This alone is enough to justify the minor disadvantages. These can all be comfortably lived with, but more maintenance costs will have to be met than with modern conventionally-roofed properties. In any event, older properties generally require much more attention than modern ones. It is worth repeating what was stated in an earlier chapter, that after about every five to seven years it is advisable to have a thatched roof inspected by a thatcher and to have any necessary cleaning and patching repair work carried out. This will, in the long term, be the most economic method of prolonging the life of the roof.

(6)

Thatched Buildings in the West Country –

Devon, Cornwall, Somerset

In this chapter on the West Country and in the succeeding chapters on the various regions of England, a review of a broad cross-section of villages and towns possessing thatched buildings is made. Villages especially highlighted contain collections of exceptionally charming or unusual thatched roofs. Some of the thatched buildings are also of historical note or are situated in villages or towns of particular interest. Many of these will therefore be worth visiting for reasons in addition to seeing the thatched roofs.

DEVON

Devon possesses a host of cob and thatched cottages. The outside walls of most are attractively whitewashed, and they form a sight well known to most West Country holidaymakers. An excellent example of a large cob and thatched village is Cheriton Fitzpaine, near Crediton. The village is set in rolling hills, and many of the thatched cottages are gathered around the fifteenth-century church. The latter is well worth a visit inside to view its three separate black and white roof sections. The long thatched building situated near the churchyard is the village school. Cheriton Fitzpaine also contains some quaint old almshouses which have an exceptionally large number of chimneys. There is another particularly good example of a cob and thatched village at nearby Thorverton. This village is also fairly large and set in hilly country. A stream runs through the village, with a pleasant bridge crossing over it. The smart thatched roofs of the cottages complete a picturesque scene, together with the pillared village butcher's shop which was built in the eighteenth century. There is a thatched country guest-house at Thorverton which is a 'listed' building.

A less beautiful but still interesting large cob and thatched village is Newton Poppleford, a few miles north-west of Sidmouth, in the Otter Valley. This possesses a variety of thatched buildings. There is an old thatched toll house situated at the top of the long main street. Nearby, several guest-houses with thatched roofs mingle with many thatched cottages. An antiques and curios shop is thatched, and also the Exeter Inn which has an unusually shaped thatched roof. About a mile to the east of Newton Poppleford is Bowd. The local inn here has a beautiful long stretch of thatch, and the ridge has been immaculately ornamented.

Otterton lies to the south, on the River Otter, and also has many cob and thatched dwellings. However, in contrast to the cob, there are also several red sandstone houses. Some of the properties standing in the village were built over three hundred years ago. The mixture of the various types of dwellings makes the village most attractive. Woodbury is situated to the west of both Otterton and Newton Poppleford in the general direction of Exeter. The village is very pretty and contains many thatched cob cottages. A beautiful view of the surrounding countryside can be obtained from Woodbury Common and also from Woodbury Castle, located just outside the village. The castle consists of a series of ancient earthworks arranged in an oval-shaped pattern.

Colaton Raleigh, near Newton Poppleford, has many good examples of cob and thatched cottages. As the name of the village may suggest, it once had a connection with Sir Walter Raleigh. It was here that he was reputed to have introduced and planted the first potatoes in England. Sir Walter Raleigh's father was once the owner of the manor house. East Budleigh, which is close by, also has many thatched cottages. In addition it has a brook, a series of foot-bridges and a very close past association with Raleigh. The church here was much attended by his parents, and inside can be seen the Raleigh Pew and the family coat of arms. The house where Sir Walter Raleigh was born, in the middle of the sixteenth century, rests in a valley just a mile outside the village on the west side. The house, called Hayes Barton, is fairly large, and it is thatched. The roof has a raised ridge and possesses three end gables, as it was built in an E-type shape. This particular shape, together with the H-shape, became favoured in Elizabethan times when the popularity of houses built around courtyards declined. The Raleigh house has several chimneys intruding through the thatched roof, and the main door has a pillared porch. Just a mile north of East Budleigh are Bicton Gardens. These are beautiful Italian-style gardens which were originally designed by André Le Nôtre, who also laid out the gardens at Versailles. There are many unique collections of trees at Bicton, together with an American garden and a countryside museum. A miniature railway runs through the gardens, and this affords a good method of viewing them and the adjoining woodlands.

Branscombe lies a little further east on the other side of Sidmouth, towards Beer Head. There are many cob and thatched cottages in Branscombe, especially in the upper part of the village. Many of these are very old. The village claims several streams, and the setting is magnificent, as the place is built on steep wooded slopes which fall away, through a chalk cliff valley, to the sea. The National Trust owns much of the surrounding land, including a thatched smithy, an old-fashioned bakery and some of the thatched cottages. The church at Branscombe is both very old and interesting. It is a small cruciform church, originally dedicated to St Winefrida, who died in the seventh century. The foundations are Saxon, but these were built upon by the Normans. The thatched house directly opposite the church was originally a priest's house, dating back to the thirteenth century. The other old thatched whitewashed cottages, near the church, are typical of the general type found in Devon. The thatch eaves above the upper windows are slightly raised in gentle curves, to maximize the available light in the upstairs rooms. Some of the cottages have thatched porches to complement the thatched roofs.

Sidford is situated a couple of miles north of Sidmouth and is therefore reasonably close to Branscombe. The Blue Ball Inn at Sidford is thatched, and the roof is equipped, rather unusually, with rainwater collection-guttering. No doubt this protects the inn's customers from water drips descending on their heads from the eaves of the thatch, during and immediately after rainstorms. The inn has a wooden upstairs front section, with doors set in the middle which could be opened to take in deliveries of stock. There are many rows of thatched cottages to be seen in Sidford and also many larger thatched houses which are often fitted with thatched porches.

Fairly close to Sidford and Branscombe but to the north east is the village of Colyford. This contains many terraces of thatched cottages. There are also many larger houses adorned with neat raised ridges on their thatched roofs. Many of the bottom ridge edges have been cut with scallops and points but the occasional house displays unusually long points mixed with the scallop crescent shapes. There are also thatched farms, thatched guest houses and a thatched restaurant in Colyford.

Honiton is located a few miles to the north west of Colyford. The town gave its name to the famous Honiton pillow lace, which was made in fair quantity in the surrounding area up to the end of the nineteenth century. The main part of this small market town consists of a long Georgian street. However, there are a few thatched cottages at the west side of the town. Some of these were built in the sixteenth century, probably at the same time as the nearby chapel of the St Margaret's Almshouses. These are of interest because they were previously used as a leper hospital. Just outside Honiton, on the Axminster road, can be found a beautiful thatched house, now utilized as the Home Farm Hotel. This stands at Wilmington in several acres of its own grounds. The thatched house was constructed in the sixteenth century, and inside are several original beamed ceilings.

About four miles north west of Honiton lies the village of Broadhembury. It boasts some lovely trees and a stream by the churchyard, and nearly all the houses and cottages are beautifully furnished with thatched roofs. They represent one of the best collected specimens of quality roof thatching in

Devon. Most of the walls of the cottages are coloured white. The village also has an interesting fifteenth-century priest's house near the church, and the village pub is thought to have been the church house. About three miles south-west of Broadhembury can be found the village of Talaton. This has many pleasant thatched cottages and a thatched schoolhouse. Gittisham, a few miles away, to the south-west of Honiton, also has many thatched roofs. Although the place is situated in a hollow, a wonderful view of the surrounding countryside can be obtained from the top of the nearby hill.

Another particularly good example of quality group roof thatching can be seen in the pretty little old-world village of North Bovey on Dartmoor. There are many excellent thatched roofs in this tiny village which also possesses a pleasant village green, an ancient cross, a fifteenth-century church and an old village water pump. There are also some fine oak trees by the green. About three miles away is the village of Lustleigh which has always been esteemed for the quality of its thatchers. It would be difficult to find better examples of a group of thatched granite-walled cottages anywhere in the West Country. Surface granite from the moors is widely used in this region, in contrast to the cob walls common in most parts of Devon. However, there are also many modern houses situated in the village. One of the best views on Dartmoor can be obtained from a hill just to the west of the village. The view is known as Lustleigh Cleave and covers an expanse of magnificent granite rocks, river and wooded slopes. One of the rocks at Lustleigh Cleave is known as the Nutcracker Rock. It is said that at one time it was possible to push the rock a little way with a gentle touch to crack a nut, and the rock would then roll back to its original position.

Widecombe-in-the-Moor lies a little south-west of Lustleigh. The village contains stone and some thatch. Widecombe also has a lovely church with a granite tower. However, Widecombe is perhaps best known for its association with the song which relates the story of 'Uncle Tom Cobleigh' and his grey mare. The village has a magnificent skyline view of boulders and rocks. Buckland-in-the-Moor is situated a short distance south. This village has an attractive moor and woodland setting which makes a perfect background for its collection of fine thatched cottages built at undulating ground levels.

Nearby on Dartmoor can be found Poundsgate, which has several thatched farms and also a thatched post office. The village of Holne is located just a little to the south of Poundsgate, near magnificent woods owned by the National Trust. There are several very attractive thatched dwellings in this lovely Dartmoor village. A thatched rectory stands about a mile down the lane from the church. This was the birth-place of Charles Kingsley, who became famous not only as a preacher of Christian Socialism but also as a great writer of novels. A very old inn can be found near the church and the main cottages of the village.

On the northern edge of Dartmoor, the village of Throwleigh has many thatched cottages, with walls constructed of granite. An old thatched priest's house stands near the church and the ancient cross. Close by lies the small pleasant village of Sticklepath which also has many thatched cottages. There is, in addition, an interesting museum in the village which covers most aspects of the Devon rural and mining industries and the machinery associated with them. The adjoining village of South Zeal also possesses several thatched cottages. It was once a copper-mining village. There is a very old and unusual inn in the village which was formerly a manor house. South Tawton, adjoining the villages of Sticklepath and South Zeal, has several thatched cottages grouped by the granite-constructed church. The thatched church house has an arched stairway as a peculiar feature.

The pretty village of Drewsteighton, close by on the north of Dartmoor, consists of a pleasant arrangement of a hilltop church and many thatched cob cottages positioned around the square. A couple of miles outside the village is an ancient burial chamber known as Spinster's Rock. This was erected from four gigantic stones, three of them supporting the fourth. The arrangement forms the only remaining *cromlech* in Devon. Also on the northern side of Dartmoor, but more in the Exeter direction, rests the village of Dunsford. This contains many cob and thatched cottages and also a medieval church. From the top of the church tower there is a magnificent view not only of the thatched roofs but also of the surrounding countryside. A well-known beauty spot lies just to the south of the village where the River Teign flows and where there is a spectacular gorge.

A short way to the east is Dunchideock. This charming place has an old post office with a thatched roof and verandah. Wooden pillars support the overhanging thatched roof. The village is rather strung out, and it is overlooked by a most unusual tall tower called Lawrence Castle. The tower was built in the eighteenth century as a memorial to General Lawrence who served with the British army in India.

Broad Clyst is located about four miles to the north-east of Exeter. The village is a particularly pretty one and there are many thatched white cob cottages to be seen. There are also several historic houses in the area, which incidentally is the centre of Devon's cider industry. The mill, standing right by the River Clyst, is owned by the National Trust. Clyst Heyes, also near Exeter, overlooks the River Culme. This is a beautiful large thatched house on the riverbank, and it is of unusual architectural interest. The building was started just before the end of the fifteenth century but took about half a further century to complete. It forms a mixture of early Tudor and Elizabethan architecture, and the thatched roof has been well preserved and renovated many times.

Silverton, about six miles on the north side of Exeter, is also of note for a different historical architectural reason. The collection of thatched cob cottages in the village represents dwellings which were built at different periods over several centuries. The origins of the buildings range from the sixteenth to the nineteenth centuries. A little to the north of Silverton, resting by the side of the River Exe, is the village of Bickleigh. The many thatched whitewashed cottages make a photogenic picture alongside the river bridge and beneath the wooded hills overlooking the village. Bickleigh Castle, near the village, dates back to the twelfth century, but it was reconstructed in the seventeenth century. The castle possesses a thatched Jacobean wing. There is also a Norman chapel by the river, and this is guarded by a rather grand gatehouse.

Mamhead lies south of Exeter and about three miles north of Dawlish. Mamhead is worth a brief mention because of its lovely thatched gatehouse lodges which are positioned at the entrance to its park. A few miles away, south of Teignmouth and nestling in a deep valley, is the village of Stoke-in-Teignhead. There are many beautiful thatched cottages in the picturesque village, and the church is worth visiting. It dates back to the fourteenth century and contains a five-hundred-year-old rood screen and a mosaic-paved sanctuary.

Further south, at Cockington, there stands an exceptionally well-known and lovely thatched village. It can also be reached by an unusual two-mile ride by landau from the sea front at Torquay. The thatched roofs in this delightful place are a big attraction to tourists. Perhaps the best-known building is the thatched forge which has three pillars supporting an overhanging porch. (A replica of this forge is often seen on horse brasses.) Cockington Court, in the village, is a sixteenth-century manor house with a Georgian frontage, and part of the house has now been converted into a restaurant. There is a thatched timber lodge, with overhanging upper-storey rooms, at the entrance to the drive to the house. A village cricket square depicts a typical English scene in the grounds of Cockington Court. The inn at Cockington is unusual because it was built in the twentieth century and furnished with a thatched roof. It is rare to find a relatively modern building in Devon finished with a Norfolk reed roof. The inn has two protruding wings at each end of the main building. Each section has a hipped thatched covering with a lovely ornamental raised ridge.

Another delightful thatched inn is found at Dartington, near Totnes, just a few miles to the west of Cockington. The inn, called The Cott Inn, exhibits one very long stretch of thatch which sweeps smoothly and low at the end of the building. The roof was originally thatched with marsh water reed obtained from Dorset. The inn was established during the early part of the fourteenth century, and inside are many low timbered ceilings. Dartington is also well worth a visit to see Dartington Hall and its lovely surrounding grounds, which are very extensive. This used to be a medieval estate, but it was converted in 1925 into an arts, education and later a rural industries research centre. The gardens and parts of the hall are open to visitors.

The village of Dittisham is located roughly to the south of Dartington and Cockington, near Dartmouth. There are some very pretty thatched cottages along the road leading to the ferry in the village, which is also well known for its plum orchards. The place has a peaceful setting with good riverside views. The church at Dittisham is very old, and parts of it date back to Saxon times. Slapton can be found a few miles south of Dittisham on the coast. A most unusual long causeway road separates the sea and pebble beach from the large freshwater lake at Slapton Lea. The lake forms a nature reserve for many bird and fish species. It also provides a source of water reeds which are used locally for thatching purposes. The long pebble beach at Slapton was used by the Americans as a training ground before the D-Day invasion. The United States army later presented a memorial to the local villagers, who were evacuated to allow the training to proceed. The memorial stands on the beach. The occasional thatched cottage can be seen at Slapton in admixture with other roof types.

A short distance south is the village of Stokenham. This claims many thatched terrace cottages, again together with cottages possessing other roof types. The Coachman's Arms separates some of the rooftop mixture in the main street. Hope Cove nestles several miles to the west of Stokenham at the tip of Bigbury Bay. The bay of Hope Cove borders the hamlets of Inner Hope and Outer Hope. The former hamlet possesses a few cottages with quaint old patched thatched roofs. Most of these cottages have small cobbled fronts to them, well decked with flower tubs in the summer. Fairly close to Hope is the village of Thurlestone. This has many smart thatched cottages with pretty gardens, and the village also has an exceptionally good golf course and beach. The grounds of The Old Rectory at Thurlestone are often open to the public during the summer.

Just a small distance along the coastline towards Plymouth,

the village of Bantham rests at the mouth of the Devon Avon. The Old Boathouse at Bantham, guarding the narrow channel entrance, has a picturesque thatched roof. The boathouse overlooks a quay and a small beach. The main Bantham beach is found over the headland. It is used for surfing, and the area is popular with ramblers. Scenic cliffs form a lovely background. Ringmore is another lovely village just to the west of Bantham. It contains many thatched cottages and some are unusual as they possess tall crooked chimneys. In contrast to the thatch, some of the cottages in the place have solid slate roofs, but this mixture of materials still makes a pleasant picture. A little further west, towards Plymouth, the village of Holbeton has many attractive thatched roof houses. Some display raised ridges with points, others have flat ridges. The fourteenth-century church overlooks the village, and it is worth a visit to see the unusual carvings and figures which adorn it.

On the extreme opposite side of the county, in the north of Devon, are the villages of Croyde and Georgeham. They are located just north-west of Barnstaple and both possess several quaint thatched cottages. Near the villages are a bird sanctuary and nature reserve; also fairly close are the huge sand dunes and long beaches of Braunton Burrows and Saunton Sands. The hamlet of Lee rests a little to the north towards Ilfracombe. Many visitors visit Lee by a boat service which operates from Ilfracombe. Lee has quite a famous thatched cottage, called The Old Maids of Lee. It is open to the public and contains a fine collection of old dresses. Further attractions of the hamlet are its splendid trees and fuchsia bushes.

Further to the west along the North Devon coast, near the favourite tourist but car-free village of Clovelly, can be found the hamlets of Buck's Mills and Horns Cross. The former hamlet offers an attractive collection of thatched cottages, which have a view over the predominantly rocky and shingle beach below. The coastal roads in the close vicinity pass through pleasantly wooded country. The hamlet of Horns Cross has an attractive thatched inn. The long building has a continuous stretch of thatch but a small separate thatched roof covers the entrance porch. The hostelry is called Hoops Inn, and hotel accommodation is available. An old-fashioned open coach forms an interesting feature in the garden. Another thatched hotel can be found at Milton Damerel, fifteen miles south of Horns Cross. The Woodford Bridge Hotel is a thatched fifteenth-century coaching-inn, and the River Torridge flows through its gardens.

Chittlehampton lies to the south-east of Barnstaple. The site of the village dates back to Saxon times. This spot rightly claims to be one of the most attractive in Devon. It also boasts the finest church tower in Devon, built in the early part of the sixteenth century. The church stands on one side of the large village square, and the remaining sides consist of thatched cob cottages. Many pilgrimages to the church were made in the Middle Ages, to visit the shrine there dedicated to a Celtic martyr.

About twelve miles south is the large village of Winkleigh with many thatched cob buildings; several surround the village church. In the square, The Winkleigh Hotel overlooks The King's Arms, which has a thatched roof. Part of the roof sweeps down to the top of the ground floor level. Three or four miles west of Winkleigh is the village of Iddesleigh. This has an unusual charm because of the high density of thatched roofs which meet the eye in all directions. There appear to be rows of thatched cottages everywhere. Most of them have white or brown walls. There is also an old thatched inn in the village.

Also of interest is the small town of Hatherleigh which is located a mile or two south-west of Iddesleigh. Hatherleigh is of note because it boasts a main street consisting of thatched cottages, with peculiar bulging walls and unusual chimneys. There are three rivers nearby which are popular for angling, and there are also stretches of moorland.

About five miles north-east of Winkleigh is the little town of Chumleigh, whose Barnstaple Inn has a thatched roof. It was built over three hundred years ago, and Charles I is reputed to have stayed at the inn. About six miles to the east of Chumleigh can be found the village of East Warlington. This claims a beautiful thatched rectory and nearby a thatched tithe barn, which has been utilized as the village hall. Th village of Witheridge lies a further couple of miles to the east. This contains a common but attractive arrangement of several thatched houses and a church gathered around a square.

Bondleigh is situated about two miles south-east of Winkleigh. This offers several thatched cottages which are very catching to the eye. These can be viewed on the way to Sampford Courtenay which is located about three miles south, towards Okehampton. Sampford Courtenay gained a place in history when a rebellion took place at the church house in 1549 against the introduction of the new English Prayer Book by Edward VI. Many of the villagers were hanged when the rebellion was quelled. In addition to the historical interest, Sampford Courtenay justifies a visit to see its many thatched roof cob cottages, all tastefully decorated and enhanced with well-kept gardens.

On the northern coast of Exmoor, at Lynmouth, a beautiful historic row of thatched cottages is situated along the harbour. The thatched Rising Sun Hotel also overlooks the harbour and forms part of the terrace. The main roof has an attractive ridge with points. A separate thatched porchway guards an entrance to the hotel. The terrace has a perfect setting by the sea, and the village has a towering ravine as a background. There was a terrible flood disaster here in 1952, when boulders were swept through the village by the swollen river. Over seven inches of rain fell in a few hours, and the situation was worsened by high tides. It caused the death of thirty people. The River Lyn has been flood-proofed since the tragedy, and the destroyed homes have been rebuilt.

Brendon rests in the heart of Exmoor, and it has many thatched whitewashed cottages. A bridge constructed in medieval times can also be seen at the village. It was originally built to allow pack-horses to cross the East Lyn River. Molland, surrounded by fine scenery on the southern side of Exmoor, offers a thatched inn for hospitality. It is called The London Inn, and a small zoo has been built in the garden. The interior of the church was tastefully restored in Georgian times. It is fitted with several tall, box-shaped pews. About four miles south west of Molland is the village of Bishop's

Nympton. The long sloping main street has many attractive thatched cottages. The church has a magnificent tower built in the fifteenth century.

Lifton rests in Devon, just on the opposite side of the Cornish boundary to Launceston. The village displays a few thatched roofs. It is interesting that the craft of thatching was often carried out in this locality with marsh water reeds rather than the more usual combed wheat reed.

CORNWALL

The number of thatched buildings found in Cornwall is very small compared to Devon's. However, there are several attractive ones to be discovered, some steeped in history. Most of the cottage walls in Cornwall are built from the local stone or cob.

Veryan, south-east of Truro by the Roseland peninsula, contains several thatched buildings of different types. The most famous are the Round Houses which have hemispherical-shaped roofs. There are five of these, two at each end of the village and one in the centre. They are mostly painted white, and each bears a cross on the top. The thatched roofs of the pair at the top end of the village differ slightly. The crosses fit on the tops of the thatched roofs through the support of metal caps. There is no obvious ridge on one of the houses, but the other possesses an apparent small ridge layer of thatch material. The pair of Round Houses at the other end of the village have thatched porches.

There is a thatched restaurant near this latter pair of Round Houses, and opposite is a house with an old but good example of long straw thatching. There is a mixture of long straw and reed thatching to be seen in the village. The church at Veryan has a small but pretty garden lake by its side. Towards the thatched Round Houses at the top end of the village is another restaurant, The Toby Jug, where both accommodation and cream teas are available. This restaurant is not thatched, but it is overlooked by a thatched house.

St Mawes is fairly nearby, in Roseland. It claims the reputation of being the warmest place in England, and it has a beautiful sheltered position. It has some well-kept and delightful thatched houses overlooking the water's edge of the bay. Some of these houses are quite large, detached and set in their own grounds. A closer look at their thatch can be taken along the hilly road on the other side of the bay leading to St Mawes Castle. A few of these have thatched porches bordering the pavement. The walls of one of the houses are completely but attractively covered with ivy, well trimmed so that it does not intrude onto the thatched roof. The castle at St Mawes is well preserved, and it was built by Henry VIII as a military fortress. Across the water, Pendennis Castle can also be seen, and the two castles guard the entrance to the Fal Estuary. There is a regular ferry boat service from St Mawes to Falmouth. During the journey across the bay, both castles can be seen, as well as the thatched houses along the sea front of St Mawes.

It is also possible to take a boat trip to visit the old thatched Smuggler's Cottage at Tolverne to the north of St Mawes. The cottage is very pretty and rests at the water's edge of King Harry's Reach. This is an extension of the Carrick Roads, across the waters of which a car-ferry service operates from King Harry's. The roof of the Smuggler's Cottage at Tolverne displays a raised ridge, cut with points along its bottom edge. The cottage has a thatched porch and also a pretty thatched canopy covering the water well in the garden. The cottage is now a restaurant. It is well known locally for its excellent Cornish teas and its beautiful waterside position. It has also been featured on television.

The Carrick Roads are a very wide and deep stretch of water, but they can be easily crossed by car ferry to visit the village of Feock. This has a very pleasant position, as it is bordered not only by the Carrick Roads but also by Restronguet Creek. Feock lies about five miles south of Truro, near National Trust land and gardens. There is an old Quaker meeting-house situated at Feock in a secluded lane by the green and amid trees. It was built in 1710 and has a steep thatched roof. The house was originally very small, with dimensions of twenty feet by twenty-seven feet, but it has since been extended. There are about four hundred such Quaker meeting-houses in England, but the Cornish one at Feock is believed to be the oldest. The thatched roof protects the house, which is constructed of cob walls. The building has green shuttered casement windows. The village church at Feock is also interesting, as the tower and belfry are separate from the main church. Although this appears somewhat unusual, it is found in other parts of Cornwall. There are two thatched cottages opposite the tower and belfry.

Gweek dominates the end reach of the Helford tidal stretch of water and lies south west of Falmouth. An unusual-shaped octagonal thatched roof can be seen on a house standing at the corner of the main street. Gweek also has a nearby seal sanctuary, and the many quiet creeks in the vicinity are ideal for bird spotters. Many herons can be seen. Just five miles to the east of Gweek can be found Helford which is reputed to be one of the prettiest villages in England. It displays a large number of palms and beautiful flower beds. There are also many white-faced thatched cottages built on each side of a very narrow creek. There are a few houses with slates on their roofs which make a contrast to the thatch. Steep hills rise on either side of the village to complete a beautiful scenic picture.

Further south, in The Lizard, the coastal village of Coverack faces towards the east. The village was formerly a smuggling stronghold. Many of the houses have thatched roofs, due to the sheltered protection the village achieves against the prevailing west winds. It is interesting to note that thatch is rarely encountered in Cornwall in village locations facing westwards and exposed to severe weather. It is therefore never necessary to employ weighted ropes to hold the thatched roofs down as is done in certain coastal areas of Ireland and the Isle of Man.

Cadgwith Cove also enjoys a sheltered position, and many thatched stone cottages abound here. The thatched roofs all appear to be at different levels, due to the hilly nature of the streets. The thatch in the village is typical reed. The cove at Cadgwith is tiny but one of the most picturesque. (Incidentally, this particular fishing village is an excellent place to buy lobsters and crabs.) Just south of the village a magnificent cliff-top chasm, called the Devil's Frying Pan, can be viewed. The chasm was originally formed when a sea cave collapsed. The approach to Cadgwith is down a very steep and narrow lane. The village of Ruan Minor sits at the top, and it also has a few thatched houses. The church at Ruan Minor has an ivy-covered tower, which was added in the fifteenth century to enlarge the original very small church. Cadgwith Cove and Ruan Minor are both close to the Lizard Point, where the famous lighthouse can be visited. Very near to Cadgwith Cove and the lighthouse is another minute inlet called Church Cove. There are some pretty thatched cottages in its immediate vicinity and also some farmyards, to remind one that there is agriculture in this apparently bleak area of The Lizard. A serpentine rock souvenir shop also exists at Church Cove, but there are a great number of these to be found at Lizard Point.

On the northern side of Cornwall, the old-world village of Crantock is sited south of the River Gannel. (The famous surfing resort of Newquay is just four miles away.) Crantock was renowned for its smugglers in the past. Many of these were reputed to have frequented the local inn, The Old Albion, and to have stored their booty there. The inn has a thatched roof and a raised ridge with undercut edge points. The ridge also has a fairly common ornamentation of horizontal liggers and cross rods. The Old Albion is four hundred years old. The church at Crantock, near the inn, has the old village stocks on display in the churchyard. The last man placed in them in 1817 was reputed to be a smuggler's son. Trebellan is located about two miles south of Crantock and also claims a past tradition of smuggling. The Smuggler's Den at Trebellan now operates as a thatched farmhouse restaurant. The quaint building encloses a small courtyard on three sides, so the thatched roof exists in three main sections. A chimney stack stands at each side of the two gable ends of the thatch.

Further eastwards along the northern coast is the popular but small, sandy seaside resort of Bude. Just a couple of miles away towards the Devon border is the little town of Stratton. This hillside town contains some quaint old thatched properties, which are situated along its narrow streets. There is also an unusual inn in the town called The Tree Inn, which was originally a manor house. A one-time occupant of the house led the Royalists to victory at nearby Stamford Hill in 1643.

The pretty village of Poughill adjoins Stamford Hill and rests a mile or two to the north of Bude. The village contains several thatched cottages and delightful gardens. The church at Poughill is interesting because it has two large medieval wall-paintings. The present church building dates back to the end of the fourteenth century. The village of Marhamchurch is also very close to Bude. This delightful place displays many thatched cottages surrounding the large square.

SOMERSET

There is a host of thatched buildings in Somerset, and the walls of quite a few are constructed of the local yellow Ham stone, quarried from Ham Hill and cut into suitable-sized building blocks. Red sandstone, obtained from the Quantocks, is also often utilized. There are, in addition, cottages constructed of cob. In the neighbourhood of Minehead, on the northern coast of Somerset, there are several collections of beautiful thatched cottages standing in the various villages and hamlets.

Minehead, in its higher town area, possesses some delightful red and yellow cottages with thatched roofs. Many of these line the steep steps road leading to the large Church of St Michael. The steps are narrow and surfaced with tiles and cobble stones. The chimneys of the old thatched cottages are tall, and they stand at the side of the cottages, intruding a little into the street. There are many strange-shaped cottages in the area, and the view looking back over the winding way to the top of the hill is very picturesque.

Selworthy, situated about four miles west of Minehead, is famed for its beautiful sheltered setting, and most of the houses in the village are thatched with combed wheat reed. A particularly lovely group of seven old thatched cottages can be found by the little green, near the stream and surrounded by trees. The thatched cottages have latticed windows and also thatched gable porches. Large chimney stacks stand and bulge to ground level at the sides of the cottages. This type of chimney stack arrangement is common in this region of Somerset. They were often used in association with bread ovens inside the cottages. The white-painted church at Selworthy is unusual, and a magnificent uninterrupted view over Exmoor can be obtained from just outside the church. Another building of interest in the village is a fourteenth-century tithe barn which is sited at the rectory.

Tivington lies a couple of miles to the south-east of Selworthy. There is a fifteenth-century restored thatched chapel in the village. A cottage, which was originally a priest's house, adjoins the chapel, and it also shares the same thatched roof. Allerford, a mile and a half to the west of Selworthy, has some pretty redstone thatched cottages, with round outside bread-oven chimneys. There is also a thatched village schoolhouse. These are gathered near an old two-arched packhorse bridge and a beautiful walnut tree. There are many such trees at Bossington which is located about a mile away. The cottages here are also mainly thatched, and their upper windows are tiny. Lynch separates Allerford from Bossington, and it has a quaint group of thatched cottages with rounded chimneys. These stand near the old mill and also close to the manor house and a chapel constructed in the sixteenth century.

The little hamlet of Horner is located about one mile south of Allerford. There is an attractive group of thatched cottages in the hamlet which is overlooked by a beautiful woodland. A small packhorse bridge spans the stream at the bottom of the wooded cleft. The village of Luccombe is a mile away, and this also has several thatched cottages near the churchyard. Most of the thatched cottages are white walled, and some again show evidence of the presence of old bread ovens.

Porlock rests between Exmoor and the sea, just a little further to the west. Porlock is a large village, and it has an excellent location for touring and horse-riding. It also possesses a good collection of thatched cottages near the church. The walls of most of the thatched cottages are either stone or cob type. The chimneys protrude from the sides of the cottages. The spire on the parish church at Porlock is unusual in that it is made of wood and is covered with shingle. The streets of Porlock are very twisty, and there is a rather notorious steep hill to test the motorist. The quaint Ship Inn in the High Street claims fame from the fact that the poet Southey was once a customer and composed, when seated there, a sonnet on Porlock.

The village of Winsford borders the River Exe to the south of Porlock and Minehead. Several attractive thatched cottages are to be seen in the centre of the village. The fifteenth-century Royal Oak Inn also has a thatched roof. This long straw thatch sweeps with a good overhang to protect the upper-storey window, built out from the front main wall. Near the inn is an old packhorse bridge, and fishing is available in the locality. Winsford's position in the Exmoor National Park ensures a beautiful rugged moorland view.

The town of Dunster, about three miles south-east of Minehead, boasts an interesting and odd collection of medieval buildings, including a castle, an old mill and a tithe barn. The castle, built during the eleventh century on the top of a hill, has been continuously inhabited since that time. An intriguing folly dominates the other end of the High Street from the castle. This is known as Conygar Tower and was built during the eighteenth century on the top of a tall mound. There is also a pretty group of thatched cottages near the old packhorse bridge. Many unusual-shaped buildings abound in Dunster, but perhaps the most peculiar is the one on the southern corner of the High Street, with its overhanging storeys and tiled roof. It was formerly a priory guest-house. There is also an old yarn market in Dunster, and nearby an ancient inn.

A little to the east of Dunster and bordering the coast can be found the village of East Quantoxhead. There are several thatched cottages in the village and a pleasant church and pond. The beach can be easily reached by walking from the village. A further eight miles to the east lies the charming village of Stogursey. The main street displays a mixture of several types of dwellings, including Georgian houses and also some neat thatched cottages. The church at Stogursey is very ancient, the Normans having built on Anglo-Saxon foundations. The remains of a priory building which stood near the church now constitute part of a reconstructed dovecot of the neighbouring farm. The ruins of a Norman castle are a short distance away. The village of Crowcombe nestles at the base of the Quantocks, a few miles to the south of East Quantoxhead. There are many good examples of thatched cottages in this pleasant place. A red sandstone church also stands in the village, together with a fine example of a church house which dates back to 1515. A thirteenth- or fourteenth-century stone market cross can also be found in the centre of the village.

The old mining centre of Priddy rests in the Mendips to the north of Wookey Hole. An annual sheep fair has been held in August at Priddy for many centuries. A thatched shelter protects the old traditional sheep hurdles which are always stored on the Priddy village green. It was believed locally that the continuous presence of the thatched hurdle stack ensured that the village would continue to hold the fair.

Mells, about four miles south-east of Radstock, can claim to be the most lovely village in Somerset. It contains many stone-walled thatched cottages, mingling among small greens and trees. Many of the cottages are built with yellow stone, which makes an attractive contrast to the grey stone employed for some of the others. There is an Elizabethan gabled manor house in the village, which was formerly occupied by a member of the Horner family. One of the family was thought to be the composer of the famous nursery rhyme 'Little Jack Horner' in the late eighteenth century. The church at Mells contains a Horner Chapel and several mementoes of the Horner family.

Taunton, the county town, has on its east road an interesting thatched building, surrounded by lawns, which was once used as a leper hospital. There is another unusual thatched house in Taunton at East Reach, which was originally built in 1500 as an almshouse, but it is no longer used as such.

Walton, about three miles south-west of Glastonbury, is a village which contains a thatched rectory, built in the fifteenth century. Another three miles to the south-west, a windmill dominates the scene at High Ham. A small thatched canopy roof covers the top of the mill, which is still fitted with a complete set of sails. A thatched late medieval priest's house stands at Muchelney, south of High Ham, in the direction of Crewkerne. The house was originally occupied by the secular priests, and it is now owned by the National Trust. It was built in the fifteenth century with stone obtained from the local quarries and has suffered little alteration since that time, other than the later construction of an upper floor dividing the great hall. The building is of a low height, and it has beautiful stone mullioned windows. Several chimney pots on the roof break through the line of the thatched ridge which is simply ornamented with cross rods and liggers. Many of the houses in the village also have mullioned windows, and it is thought that these were originally salvaged from the ruins of the old abbey.

The village of Pitney is about three miles away to the north east. Many thatched stone-walled cottages gather around the church in this pretty spot. The small town of Martock is situated to the south. This interesting place contains many old thatched houses constructed of stone. There are also many other old buildings in the town, and some date back to the fourteenth century.

Ilminster lies to the south-west of Martock, about six miles from Crewkerne. A fine row of typical West Country whitewashed thatched cottages can be found fairly near the colonnaded market house in the square. The white cottages make a refreshing contrast to the many yellowish Ham stone buildings in the immediate vicinity. The village of Stocklinch, a couple of miles away, contains some more Ham stone buildings. In this instance, many are pretty cottages topped with thatched roofs. An old church stands among them, and there is also, rather unusually, a second church in the village. An old thatched farmhouse can be found approximately a mile from Stocklinch at Ilford Bridges. It is of interest because it is thought that it may have been once used by Judge Jeffries as a temporary court.

The village of Hinton St George is situated about four miles from Ilminster and about the same distance from Crewkerne. It is a beautiful village centred around a wide main street.

There is an old thatched four-gabled Priory Cottage in the village, together with many Tudor, Jacobean and Georgian and a few Victorian buildings. Many of the houses are constructed of Ham stone. The thatched Priory Cottage is located opposite the medieval cross in the main street. An unusual ceremony takes place in the village each year, on the last Thursday of October. The clocks have then normally changed, and it becomes dark early enough for the 'Punky' celebration to be held. This is a warm-up for Hallowe'en, and the children in the village have a procession through the streets, with lanterns made from turnips and mangolds. The ceremony is unique to Hinton St George.

The large town of Yeovil is located just a few miles to the east of Hinton St George. Yeovil offers a magnificent beauty spot known as Nine Springs. (Its position is at the foot of Henford Hill.) The name originates from nine separate springs which feed a lake surrounded by beautiful trees, shrubs and many small paths. A delightful rustic thatched cottage overlooks the scene. In addition to the main thatched roof, the cottage has a separate stretch of thatch covering a timbered verandah.

Above: Restored cottages at Cavendish, Suffolk.

Below: A thatched cottage at Elsworth, Cambridge.

Above: The Smith's Arms at Godmanstone, Dorset – one of the smallest pubs in England.

Below: Cockington, near Torquay, South Devon. (A thatched forge at left).

Above: The Fox and Hounds Inn at Elsworth, Cambridgeshire. (A thatched bus-shelter beyond).

Below: The Royal Oak Inn at Winsford, on Exmoor (Somerset).

Above: A yeoman's house at Bignor, Sussex.

Below: The village street, Cropthorne, Warwickshire.

Opposite: Mill House, Raby, Cheshire.

MILL HOUSE

Above left: A thatched inn-sign at Fen Drayton, Cambridgeshire. *Above right:* The Umbrella House, Lyme Regis, Dorset. *Below left:* A thatched Dutch octagonal cottage, dated 1621, at Canvey Island, Essex. *Below right:* One of four Round Houses at Veryan, Cornwall.

Above: A thatched Dutch octagonal cottage at Canvey Island, Essex.

Below: An angler's hut on the River Test at Longstock, Hampshire.

Above: The charming village of Chelsworth, Suffolk.

Below: Wherwell, Hampshire.

(7)

Thatched Buildings in East Anglia –

Norfolk, Suffolk, Essex, Cambridge

The counties of East Anglia contain a large number of thatched buildings. When they were built, the huge growths of Norfolk reed and sedge ensured the ready availability of roof constructional materials. The introduction of pantiles, manufactured from the local clay, offered an additional choice in the eighteenth century. The walls of the buildings were also built from locally available products. Many of the thatched cottage walls were constructed of brick and flint. The bricks were originally made from the local clay and the flints obtained from the chalk deposits in the region. The clay bricks or lumps were unfired, as it was not until the seventeenth century that the modern type of fired brick became widely available in England. Many houses were also built of timber, due to the vast expanses of woodlands in the area which yielded an unlimited supply of wood.

Many houses were also constructed in the timber-framed style. It is of interest that the introduction of timber-framed houses in Elizabethan times was to allow the weight of the roof to be more easily carried and transmitted to the ground. This is not so important with a thatched as with a heavy tiled roof. However, the use of a timber frame allowed the walls to be in-filled between the timber boxes, with a variety of materials, such as wattle and daub. Flint and chalk mixtures were often employed as the in-filling in East Anglia. In the late seventeenth century it became a more common practice to use bricks as the in-filling throughout many regions of England, but it was not found necessary to bond them together because the timber frame construction carried the main load. The bricks were often arranged in a herring-bone pattern to obtain water-tight joints.

NORFOLK

Norwich forms a good centre for the Norfolk Broads, and Salhouse Broad lies to the north-east of the city. The area provides many acres of reed swamps which yield material suitable for thatching purposes. The village of Salhouse overlooks the lovely waters of the Broad. This picturesque place has All Saints Church, with a thatched roof and an arcade which dates back to the fourteenth century. The church also displays a small sanctus bell which is attached to the rood screen. (It is thought that only one other example of such a bell exists.) It is interesting to recall that, during the Middle Ages, thatch was first introduced as a temporary roof-covering for churches. Today there are still many churches scattered throughout the countryside with thatched roofs, and the majority of these are situated in East Anglia. In addition to the church, there are also some beautiful Norfolk reed thatched cottages to be seen at Salhouse. Several of the roofs are finished with delightful and richly ornamented ridges.

Horning, the popular Broadland yachting and angling centre, is very close to Salhouse. Ye Olde Ferry Inn forms a well-known and friendly visiting place, near the Horning ferry. The inn has a magnificent Norfolk reed thatched roof. The thatching gives the roof an unusual two-stepped layer effect. This was done by the fixing of a wide apron of thatch material below the raised ridge. The apron ends about halfway down the roof. The bottom edge of the thatch apron layer on the main thatched roof section is cut in line with the bottoms of the dormer windows. The windows possess small separate ornamental aprons below them. The ridge of the roof has scallops and points, and it is beautifully ornamented. These imitation scallops or crescent shapes, alternating with points, are an exceptionally common feature found cut in thatch ridges in East Anglia, as a bottom decorative edging. A further stretch of thatch, in the shape of half a sectioned cone, protects some circular bay windows abutting from the wall of the inn, under a gable end of the thatch. There are several other interesting and charming thatched buildings in Horning. Typical of these is the thatched barn which stands at Horning Hall, about a mile and a half to the east of the church. This thatched building was formerly used as the chapel for St James's Hospital which no longer exists.

The neat and tidy little village of Woodbastwick sits between Salhouse and Horning. This contains some delightful thatched houses, although there are some pantiled roofed properties as well. The pretty village also possesses a thatched shelter on the green and a small thatched church. The village of Coltishall rests by the River Bure, just to the north-west of Salhouse. The village has attractive surrounding scenery, including some lovely pine trees. The Church of St John the Baptist has a continuous thatched roof, covering both chancel and nave. The church is rather unusual in that it contains some windows which date from Saxon times. Just outside the village, another church can be found, but this decays in a relative state of ruin.

Another quaint thatched church can be discovered in the tiny village of Hoveton St Peter, which lies about three miles to the east of Coltishall. The church, like the village, is rather small, and its total length approximates to only forty feet, with a width of about eighteen feet. The hamlet of Irstead is a further three miles to the east. The church here claims a thatched roof, and the exposed underside of the thatch can be viewed from within the church.

The popular yachting centre of Potter Heigham is a small distance to the east of Irstead, on the banks of the River Thurne. The Church of St Nicholas in the old village boasts a thatched roof over the nave and chancel. It also has a round Norman tower. In addition, the village claims a rather unique three-arched bridge which dates back to the thirteenth century. Horsey Mere is situated a short way to the north-east of Potter Heigham. The water in this particular Broad is very brackish, as it is separated from the North Sea only by a large sand-dune barrier. Many unusual birds and other wild life can be found in its locality, due to the saline nature of the water. The church at Horsey is of interest, and it has a thatched roof which again covers both nave and chancel. In contrast to this, the church at nearby West Somerton has a thatched roof only over the nave. This church commands a good view of the sand dunes and the sea.

Rollesby Broad lies about three miles south-east of Potter Heigham. The village of Rollesby contains some lovely old thatched cottages, some of which date back to the sixteenth century. Many equally fine thatched roofs can also be seen in the village of Stokesby which is situated another three or four miles to the south. There are also pantiled roofs alongside to compare with the thatch. The church has a thatched roof which covers both nave and chancel in a continuous stretch. In variation to this, the fourteenth-century church at the nearby village of Beighton can claim only a thatched nave.

Further to the north, on the Norfolk coast, at Paston, there is also a thatched church. This is the Church of St Margaret, and it dates from the fourteenth century. The nave roof of the church is thatched with Norfolk reed, and inside, the open timbers and beams of the roof can be viewed. Interesting aged paintings adorn the walls of the church. One of the finest examples in Norfolk of a thatched tithe barn is also situated at the village of Paston. The barn measures approximately fifty-four yards long, and the buttressed walls are constructed of flint. It was built in 1581. The massive thatched roof is supported underneath by a series of tie and hammer beams. The pretty village of Bacton nestles close to Paston. The church at Bacton is located approximately half a mile from the coast and it, like the one at Paston, has a partially thatched roof. The ruins of Bacton Abbey are also worth a visit, and these are locally known as those of Bromholm Priory.

The village of Ridlington is sited about two or three miles to the south. This possesses a lovely thatched barn made of brick, with Dutch gable ends, which form a reminder of East Anglia's former close contacts with the Netherlands. The barn stands opposite the church, which is also of interest because of the gigantic key required to open its antiquated door. The village of Crostwight is close by, and the church here has a thatched roof over its chancel. Inside there is a decorative work depicting a tree with the seven deadly sins.

Costessey Park can be found just outside Norwich, on its western perimeter. The lodge to the park was built in the late eighteenth century and was designed in the *cottage ornée* style. A beautiful conical-shaped thatched roof protects the circular lodge building. A chimney pot protrudes through the apex point of the thatch. A little further south, about six miles from Norwich, is situated the tiny village of Ketteringham. This has some lovely thatched and timbered cottages standing next to the church, which displays within many fine old brasses.

Much further south, the village of Bressingham rests near the border with Suffolk. The many beautiful thatched cottages, set in orchard scenery, make a delightful picture. There is also a thatched inn opposite the church. An unusual attraction to visitors is Bressingham Hall, which contains a remarkable collection of steam engines, tractors and fairground engines, and a miniature railway. The large grounds and gardens of the Hall are magnificent.

There is a thatched church at the village of Caston, which can be found several miles to the south-west of Norwich. This church claims an interesting history of roof changes. In medieval times, it was originally covered with a thatched roof. Later, in the middle of the nineteenth century, the decision was taken to convert the roof structure to Welsh slates. However, due to problems encountered with leaks in the roof, it has recently been changed back to thatch. Another thatched roof covers the church at the nearby village of Rockland St Peter. This small church has a Norman round tower. Great Hockham lies about four miles to the south-west of Rockland St Peter. Several old thatched cottages, together with some houses which have pantiles on their roofs, border the green.

The village of Reedham is located to the south-east of Norwich in the direction of Great Yarmouth and Lowestoft. At Reedham, the west perpendicular tower of the Church of St John the Baptist is massive, and the church is thatched. A seventy-foot-high windmill makes an interesting landmark just three miles outside the village. The windmill is one of the finest in East Anglia. It is in working order and is open to interested visitors. It is now administered by one of the government departments. At nearby Heckingham, there is another thatched church, whose roof is exceptionally neatly ridged, with decorative scallops and points. The thatched roof spans both nave and chancel. The village of Hales rests a couple of miles to the south of Heckingham. The small church here also has a thatched roof covering nave and chancel, a Norman round tower and an unusually beautiful Norman doorway.

There are several other thatched churches to be discovered in the immediate neighbourhood. Sisland lies about three miles to the west of Hales and contains a relatively modern church, built in the eighteenth century, with a thatched roof. The village of Seething nearly adjoins Sisland, and a thatched roof protects the nave of the church. Stockton is situated about two miles to the south of Hales. The church is quite small and again possesses a thatched roof. The village of Fritton snuggles very close to the Suffolk border, to the south-east of Reedham. The small church in the village has yet another thatched roof, and it also boasts a very early Norman round tower. The thatched roof is divided into two separate sections, one covering the nave and the other the chancel. An ornamented ridge adorns both thatched roof sections.

SUFFOLK

The village of Herringfleet is a couple of miles south-east of Fritton but rests in Suffolk. It also has a church with a Norman round tower and a thatched roof. However, the tower is not quite of such an early date as the one on the thatched church at Fritton. In addition, an old thatched barn still stands at Herringfleet. The barn walls are constructed of pebbles and bricks; the building was once used as a refectory for the nearby Augustinian priory, founded in the thirteenth century. Traces of the priory's former site are still evident. The unspoilt, beautiful village of Somerleyton is very close to Herringfleet, on its eastern side. The entire village was created in the *cottage ornée* style, when Somerleyton Hall was reconstructed in 1844. This spot contains many neat thatched cottages placed in a delightful setting of village green and pump. Somerleyton Hall displays magnificent grounds which surround the main house. A maze, constructed from clipped yew trees, forms an intriguing feature of the garden.

At Barsham, near Beccles, the nave roof of the Church of the Most Holy Trinity is thatched, and the church also has an early round tower. The rectory which stands close to the church is of interest because Nelson's mother was born there in 1725. The nearby village of Ringsfield also boasts a thatched roof on its church. Some delightful thatched almshouses can be viewed at Homersfield, located about seven miles to the west. The thatched roofs are unusual in that they are relatively modern. The almshouses are arranged in a horseshoe-shaped terrace, and a beautiful ornamented raised ridge caps the wide and low sweeping expanse of thatch. A thatched canopy, supported on pillars, shelters a well in the garden. Several miles to the east, on the coast near Lowestoft, can be found the village of Pakefield. The church on the cliff at Pakefield claims justifiably to be exceptional, due to its newness and roof type. It was constructed after the Second World War and yet was still furnished with a thatched roof. The original church had been destroyed during the war.

A much older but also exceptional church exists at Reydon, which is further south, near Southwold. The roof of the church appears to be completely tiled when viewed from one side. However, a closer inspection, on the other side of the church, reveals that it is thatched. This dual-material roofing was initiated in 1880. Although thatch was in fashion at that time, the clergy in this instance appeared not to think so. They deemed it better for the church to be tiled on the front, where the road passed, with only the back roof thatched where it would be less noticeable. In contrast to this reticence, a very long proud stretch of thatch can be viewed on the nave roof of the Church of St Peter at Theberton, which is also in East Suffolk. The church has a round tower and a Norman doorway. It is said that the church was not unknown to smugglers during its past history. The church at Bramfield displays a further thatched roof. The village lies to the north west of Theberton. The thatched roof is magnificently ornamented and has a ridge bottom edge of points. In addition, the roof is further decorated with a full-length apron of alternating long and short points positioned halfway down the thatched roof surface. A detached round tower constitutes a rare feature of the church. The tower stands a little distance from the main church; it has very thick walls and also some medieval bells.

There are many thatched villages and thatched cottages in farmland settings in Suffolk. John Constable, the famous landscape painter, frequently used some of them in his work. Constable, the son of a prosperous miller, spent much of his youth at Flatford Mill which is now owned by the National Trust. The mill is located on the River Stour at the southern boundary of Suffolk, by the side of a wooden bridge with the famous Willy Lott's cottage nearby. A thatched cottage rests at the foot of the bridge leading to the mill and Willy Lott's cottage. The very pretty village of Higham is about four miles away, to the north-west of Flatford Mill. It hides among lovely tree-lined slopes. An unusual thatched house adjoins Higham Post Office. The top third section of the roof, when viewed from the roadside, is thatched and also nicely ornamented with scallop shapes, liggers and cross rods. The bottom two thirds of the roof area is tiled. A rare combination of thatch and tiles is therefore incorporated into the same roof side.

The village of Whatfield lies about seven miles to the north of Higham. Delightful countryside surrounds Whatfield, and many very old thatched cottages abound here. Just about three miles away to the west can be discovered the picturesque village of Chelsworth. This possesses many thatched cottages which are made additionally attractive by the artistic timbers employed in the construction of many of the dwellings. The River Brett runs by the village and adds further charm. The village of Lindsey is about three miles south of Chelsworth and Whatfield. This contains several very old thatched buildings, and exceptionally noteworthy is the Chapel of St James, surrounded by a garden. The exact age of the building appears to be lost in antiquity but a straw-thatched roof still protects it, as it has for innumerable centuries.

An outstandingly beautiful thatched village can be explored at Cavendish which is to the west of Lindsey, a few miles south of Bury St Edmunds. The many attractive thatched timber and colour-washed cottages in this delightful place depict the popularity of this type of dwelling in the south of Suffolk. The cottages of Cavendish have been exceptionally well restored and cared for; the walls are mainly pink in colour. The thatched roofs are immaculate, and they are finished with fine pointed bottom edge ridges. Many of the roofs are of the outshot design, in which one side extends below the level of the other side – a form of construction once fashionable in Suffolk. The thatched cottages at Cavendish gather around the very large village green, near the fourteenth-century St Mary's Church. The church is unusual because it has a jutting stair turret, and the tower conceals a small room uncommonly fitted with a fire-place. The church also owns a sixteenth-century brass lectern, believed to have been donated by Elizabeth I.

A few miles away, to the north-east of Bury St Edmunds, is the village of Ixworth Thorpe which has the very small thatched Church of All Saints. The building also has an unusual wooden bell turret. One of the bench ends inside the church depicts a thatcher with a comb. The Norman doorway of the church is of additional interest in that it is rather tiny, being only three feet wide and just over five feet high. Another small Norman church with a thatched roof can be found in the village of Thornham Parva which lies a few miles to the east of Ixworth Thorpe. Inside the church, a beautiful retable can be viewed, and the paintings on it have been dated to the beginning of the fourteenth century.

There are many other thatched villages in the region of Bury St Edmunds, and typical of these is Flempton, located about four or five miles to the north-west. This place contains a church, a pleasant green and close by it a quaint row of cottages with thatched roofs. Dalham holds a position to the west of Bury St Edmunds, towards the border of Cambridge. The River Kennet runs by the village which displays a great number of thatched cottages. The spot is also of interest because of the existence there of Dalham Hall, once owned by the family of Cecil Rhodes, who played such a prominent role in the historical and political development of South Africa.

Newmarket Heath sprawls just a few miles to the west of Dalham. Newmarket forms the main centre for flat horse-racing in England and the headquarters of the Jockey Club. Many studs and racehorse-training establishments abound in the area. The racecourse claims a very attractive long thatched building. This overlooks the paddock and parade-ring area, and the thatch protects the open-fronted saddling stalls. The lengthy stretch of thatch exhibits a very neat ridge which is well ornamented with points.

ESSEX

The village of Chrishall borders the county of Cambridge and is therefore sited in the north-west corner of Essex. Chrishall contains several quaint seventeenth-century timbered cottages with thatched roofs. Also of interest in the village is the ancient earthwork mound, still surrounded by a moat. A further collection of thatched timbered cottages exists at the village of Langley, approximately three miles to the south of Chrishall. The cottages nestle in the vicinity of the church, which commands a magnificent view of the surrounding countryside. The village of Wendens Ambo lies just a little to the east near Saffron Walden. A large thatched barn dominates the approach road to the Norman church at Wendens Ambo. The expanse of thatch sweeps beautifully over the three gables of the barn.

The small town of Thaxted is situated a few miles to the south-east. The name 'Thaxted' conjures the thought that the town may be full of thatched buildings. In fact, the opposite is true, and there is only a handful of thatched roofs left in the town. However, the villages surrounding Thaxted show a host of thatched properties. The three-storeyed Guildhall in Thaxted dates from the fifteenth century, and each storey overhangs the one below, supported by wooden posts. The ground floor opens on three sides. It is of special interest because it guards some old long pole hooks which were originally made for pulling burning thatch away from roofs. The Church of St John the Baptist at Thaxted also rewards a visit as it is judged one of the finest in Essex.

The village of Birdbrook is located to the north-east of Thaxted, near the Suffolk border. Birdbrook contains many fifteenth-century buildings, some timber framed with overhanging storeys, some thatched. (One lucky new owner of an old thatched cottage in Birdbrook found a treasure trove under a flooring stone when he lifted it, after moving into the cottage during 1977. The trove consisted of about one hundred gold sovereigns from the period 1825 to 1845.) The small town of Great Bardfield is situated five miles to the east of Thaxted. An unusual thatched building stands in the High Street. This is the tiny sixteenth-century Cottage Museum which has been well renovated, with the thatched roof supported on open rafters. Despite the small size of the building, it has a comparatively large chimney stack. The museum exhibits many items associated with Essex's rural life, including a fine collection of corn dollies. Another feature of Great Bardfield is the brick tower windmill which is readily visible from most directions. The village of Wethersfield can be found about three miles to the east of Great Bardfield. It contains many old houses but of specific interest is the thatched cottage which in medieval times was used as a chapel.

Felstead, well known for its public school, is about seven miles south of Wethersfield. The sons of Oliver Cromwell were just some of the many privileged who attended the school in the past. A thatched cottage, called Quakers Mount, stands by one of the several pleasant greens to be discovered in Felstead. The thatched cottage displays a series of decorative moulded shapes, depicting fishes, a wheatsheaf and a windmill. It is noteworthy that this type of pargeting plasterwork often composes a typical feature of East Anglian architecture. The villages of Henham and Ugley nestle close to one another, a few miles west of Thaxted. Both places are charming and well kept and contain interesting collections of thatched cottages. The church at Henham overlooks an exceptionally elegant thatched cottage, while at Ugley a quaint group of thatched cottages guards the road to the west of the church.

The village of High Roding lies about eight miles south of

Thaxted. It boasts an unusual collection of gabled and thatched cottages. Many have plastered walls. Some (erroneously) appear to be only single storeyed, due to the sweep of the thatch passing the upper windows to the ground floor ceiling level. Most of the thatched roofs also terminate with tufted points at the pinnacles. Some of the windows in the end walls of the cottages are sheltered with the thatch moulded in the Sussex hip style, others with just a gentle sweep cut into the thatch eaves layer. The tiny village of Aythorpe Roding is situated a couple of miles away. It contains some neat, well-maintained thatched buildings, including a timber-framed house constructed nearly four hundred years ago, yet still in excellent condition. The village of Beauchamp Roding is a short way to the south. This also claims several thatched cottages and a fourteenth-century church which is unusually placed in a rather isolated position, in the middle of open fields.

Further south, Chadwell St Mary can be found a couple of miles from Tilbury. There are many new houses here, but there is still a quaint fifteenth-century timber-framed house with a thatched roof standing at the crossroads. Hordon-on-the-Hill lies a little to the north and contains a picturesque collection of thatched dwellings. There are also many old buildings in the village, including a timbered inn dating back to approximately the fifteenth century. The village of Fobbing is situated about three miles to the east and also claims a five-hundred-year-old inn. This village offers a contrast in roof styles, with thatched cottages in admixture with tiled houses. Some of the thatched cottages are constructed in the Wealden shape. In this type, the building is constructed around a central hall, and the roof sweeps continuously over the whole, including any overhanging storey.

Canvey Island is set in the Thames estuary, just a few miles away. The island was originally saved from the river by the employment of a Dutch engineer, called Joos Croppenburgh, as long ago as 1622. He built a complete sea wall around the island because it was subject to flooding by the spring tides. For his labours, he received one third of the island as a reward. Two thatched octagonal-shaped Dutch cottages, built in 1618 and 1621, still remain on the island. One of these quaint restored thatched cottages was opened as a small museum in 1962. Octagonal-shaped thatched buildings can be found in other parts of Essex. The tiny village of Little Bentley, in the north-east of the country near Colchester, yields such a building. In this case, it is a thatched octagonal lodge which can be found near the churchyard.

CAMBRIDGE

The county of Cambridge claims a host of thatched dwellings, and the highest density of these is found in the south, gathered fairly near to the city of Cambridge. Many of the thatched roofs are constructed of long straw but there are also many reed roofs. St Benet's Church, in the city of Cambridge, constitutes the oldest building to be found in the county. It dates back approximately one thousand years and displays a Saxon tower. The church exhibits several historical items, including an old iron hook, once utilized for pulling down burning thatch.

The village of Fen Ditton nearly adjoins the eastern suburbs of the city. The village affords a well-known vantage point, at Ditton Corner, for viewing the Cambridge Eights Races held on the river during May. For this reason, the thatched cottages in the village are a familiar sight to the many sporting visitors. A little further eastward, the village of Lode is just balanced on the edge of the Fens. This place contains a picturesque group of thatched cottages and also a thatched village hall. Another feature of renown in the village is Anglesey Abbey, now administered by the National Trust and open to the public. The Abbey was founded in 1236, but it was converted into a manor at the end of the sixteenth century. The house has also been modified with several recent alterations, and it guards many art treasures. The gardens of the house are very large and also beautiful.

The village of Great Wilbraham lies a short step further to the east of Cambridge. This village has many thatched cottages, several of which have hollyhocks and other attractive cottage-garden flowers to enhance the charm of their thatched roofs. The village of Kirtling rests near the Suffolk border, to the east of Great Wilbraham, offering some fine examples of long barns with thatched roofs, standing alongside the village pond. There was once a magnificent Tudor mansion at Kirtling, where Queen Elizabeth I was lavishly entertained. It is now in ruins, but a new house exists on the site which still retains the original gateway of the former Tudor mansion. This house, which is moated, can be found near the Roman Catholic Church. The hamlet of Upend adjoins the village of Kirtling, and this warrants a visit to see its very lovely and richly thatched cottages.

The small village of Grantchester is about two miles south of the city of Cambridge. The name is well known to literary scholars through the poem written about Grantchester by Rupert Brooke, who once lived there. The Red Lion Inn in the village has a thatched roof. A good collection of thatched cottage roofs exists near the Church of St Andrew and St Mary. Further south from Cambridge, the village of Barrington reclines on a slope by the River Cam or Rhee. This pleasant spot has a thatched inn, the Royal Oak, which is a timbered building. Several of the cottages in the vicinity are also thatched, and they border, along with orchards, the very large village green. The village of Foxton is only a mile away. This claims many thatched cottages, several of which have plastered walls, which contrast with the modern brick also found in the village and the flint originally used to build the old church.

The village of Pampisford is located about five miles to the east of Foxton and produces several interesting thatched buildings, including the late-sixteenth-century Chequers Inn. Many thatched cottages also line the village street, and the place is made additionally pleasant by the many beautiful trees which encompass it. Another building of note in the village is the timber-fronted post office, which stands opposite the thatched inn. The remnants of an ancient earthwork defensive ditch, known as Brent Ditch, can also be found just outside the village.

Many orchards surround the village of Meldreth, which is situated about three miles to the south-west of Foxton.

Meldreth has several attractive thatched houses. The heads of two huge fire hooks adorn one of the inside walls of the church. These were once attached to long poles for pulling away thatch from burning roofs. The small village green displays a whipping-post, old stocks and the base of an ancient cross. The tiny village of Abington Pigotts is four miles away to the west. Many picturesque thatched cottages contrast with tiled ones along the village street. The beauty of the village is further enhanced by the many trees and orchards.

Hinxton village lies to the east of Meldreth near the Essex border. It vaunts several pretty thatched dwellings, but there is also an unusual mixture of timber-framed and Victorian yellow brick cottages in the village. The occasional timbered house has an overhanging storey. The ancient neighbouring place of Ickleton also yields many thatched cottages, and some of these are constructed of yellow brick. There are several other interesting old houses to be viewed in the village. Rare medieval carvings decorate the coping of the churchyard wall. These include a crocodile and a fox. The village of Great Chishill rests further south, near the borders of Hertfordshire and Essex. An exceptionally picturesque group of thatched cottages stands opposite the church, and an old water pump is still in evidence outside one of the cottages. Just a mile from the village, an old white windmill complete with sails can be discovered.

The attractive large village of Comberton is situated about four miles west of the city of Cambridge. It claims many old buildings, including a collection of quaint thatched cottages by the brook. About three miles further west, the village of Bourn has several thatched barns in its immediate vicinity. It also boasts the oldest windmill in England, dating back to 1636. It is still in working order, and visitors can be shown around. The windmill is of the post-mill type, in which the sails, body and machinery revolve around a central post to face the current wind direction.

The village of Elsworth can be found about four miles to the north of Bourn and about eight miles to the west of the city of Cambridge. Many pretty thatched cottages line the streets of the village, which also has a thatched bus-shelter. A stream winds its way among the cottages and under several rustic timber bridges. The place also has an Elizabethan house, formerly used as the Guildhall, together with a manor house of the same period erected alongside the village green. The village of Eltisley lies a small distance to the south-west and is composed of many attractive thatched cottages, gathered around the large village green and church. It is interesting to note that a sister of Oliver Cromwell was married in the church in 1636. Her husband formerly lived in the village. Croxton is two miles to the west of Eltisley, in Croxton Park. Several thatched cottages stand in the village, and most are weather boarded. An attractive timbered house adjoins the thatched cottages. The nearby church and lake make a lovely background setting for the thatched roofs.

Fairly close by, to the south, rests Great Gransden, which was formerly in Huntingdonshire but is now in Cambridge. This large, lovely village contains many timber-framed thatched cottages, most of them with white or pink plastered walls. A few miles to the north can be found the Roman town of Godmanchester which was also formerly in Huntingdonshire but is now in the county of Cambridge. An unusual collection of old thatched cottages constructed of brick and timber can be viewed in the town. Many examples of other types of timber-framed houses, built in the sixteenth and seventeenth centuries, also exist in the town.

The village of Brampton lies just to the west of Godmanchester. There are many quaint old cottages to be viewed in the village. A few possess beautifully thatched long straw roofs. Other buildings of interest in Brampton include an old watermill, a three-hundred-year-old inn and a farmhouse where Samuel Pepys resided. An unusual obelisk signpost stands on the village green. Good examples of thatched roofs, brick and timber-framed cottages can be studied in the village of Hemingford Grey which is located just a little to the east of Godmanchester. One of the timber-framed cottages in the village dates from 1583. This former Huntingdonshire village commands beautiful views from the banks of the River Ouse, and the church stands on the bend of the river. There is also a twelfth-century manor house in the village, which is surrounded on three sides by a moat. The river runs along the remaining side of the grounds of the house. Hemingford Abbots neighbours Hemingford Grey and contains many immaculately thatched dwellings. It also offers a thatched inn, called The Axe and Compass. An elegant ridge, with scallops and points, adorns the top of the roof.

The village of Fen Drayton is situated fairly close to Hemingford Grey. This displays many interesting thatched buildings, some of which are timber framed. One house of historical note stands opposite the church. This thatched house has a motto written in Dutch over the door; it translates to 'Nothing without Labour'. It is believed that the Dutch engineer Vermuyden, who drained the Fens and reclaimed thousands of acres of waste land during the seventeenth century, once lived in the house. The Three Tuns Inn at Fen Drayton overlooks the village stream. This building has a beautiful thatched roof section, richly ornamented with scallops and points. An adjacent section of the inn is timber framed and tiled. The inn shows an attractive sign outside, mounted on a post and protected by a small, quaint, conical-shaped canopy of thatch. The inn probably dates back to the fourteenth century, and inside are some magnificent carved oak beams highlighting the ceiling. The village of Long Stanton is about four miles to the south east of Fen Drayton. An attractive group of thatched cottages stand by the Church of St Michael, which has a low sweeping thatched nave roof. There is another church in Long Stanton which is located less than a mile away. However, this Church of All Saints does not possess a thatched roof.

The village of Landbeach lies about four miles to the east of Long Stanton and about the same distance north of the city of Cambridge. This spot contains some thatched and timber-framed houses. Several thatched barns are also to be seen in the surrounding fields. Just a couple of miles away to the south is the village of Milton, which still retains a few old thatched cottages, despite the building of many modern houses in the locality. One of the more interesting thatched buildings is known as Queen Ann's Lodge. In former times, this lodge was an inn, and it displays a collection of old medallions on its walls.

The medium-sized town of March sprawls in the northern part of the county of Cambridge. The town is industrially well known for its large railway marshalling yard. It is also much visited to view the Church of St Wendreda which has a magnificent double hammer beam constructed interior roof, one of the best examples to be seen in East Anglia. There are some noteworthy old thatched buildings in the town, and The Ship Inn, with a thatched roof, dates back to Georgian times. Ramsey can be found about ten miles away to the south-west of March. A few single-storey thatched dwellings, constructed of yellow brick, remain in the town. Most possess tall brick chimneys and shuttered windows. An unusual feature of the thatched houses consists of the outside hatches in the gable walls, which allow access to the attics under the thatched roofs. When the houses were constructed, provision was made for entrance to be gained to the attics from the outside and not the interior. Other items of interest in Ramsey are the remnants of the fifteenth-century gatehouse, formerly part of Ramsey Abbey, which was founded in the tenth century. This is now administered by the National Trust, and it is open to the public. Seven miles to the north-west, the large village of Yaxley settles close to Peterborough. A thatched inn stands in the centre of Yaxley, and there are also several delightful black-and-white painted thatched cottages nearby.

The village of Wicken stands in the eastern region of the county, two miles south-west of Soham. Wicken Fen was much used at one time for sedge production. The sedge constituted a much-utilized local fenland thatch material. The fen at Wicken is preserved by the National Trust and is kept as similar to its original state as possible. It is now a nature reserve. Sections of the fen are open to the public, and these are of special delight to all who enjoy walking, viewing the huge growths of sedge and studying bird life. A restored windmill reminds visitors of the methods originally utilized to drain the fens and reclaim the land by pumping the water into the dykes. The Coach and Horses Inn in the village is covered with a thatched roof. Spinney farmhouse just outside the village is of historic interest because one of Oliver Cromwell's sons once lived there. Burwell Fen lies a little to the south of Wicken, and the manor house in the village has several thatched stone barns. A windmill, in its original form, can also be viewed near the village.

(8)

Thatched Buildings in the South and South-East –

Dorset, Wiltshire, Avon, Hampshire, the Isle of Wight, Berkshire, Greater London and Surrey, Sussex, Kent

DORSET

A host of thatched cottages and various other thatched buildings can be found in Dorset. Most of the thatched roofs consist of combed wheat reed, although some are also thatched with local marsh water reeds. Long straw thatching can also be seen, particularly in the south-east region of the county. Norfolk reed is less commonly encountered. The county town of Dorchester forms a good centre for touring Dorset. A thatched building of historic note in Dorchester itself is the Hangman's Cottage, standing by the river running alongside the prison. Here the executioner used to live.

Other interesting thatched properties on the outskirts of Dorchester include the cottage at Higher Bockhampton where Thomas Hardy, the novelist, was born in 1840, and the former residence of William Barnes, the dialect poet, situated at Winterborne Came. The interior of the Thomas Hardy cottage can be viewed by appointment with the tenant, as it is now National Trust property. The cottage gardens are open to the public without the need for an appointment during the summer months, so the exterior of the cottage can be readily viewed. The churchyard at Stinsford, about three miles away, is also worth a visit as the heart of Thomas Hardy was buried there in 1928, although his ashes rest in Westminster Abbey.

At Winterborne Came, the former William Barnes residence consists of a beautiful thatched rectory, surrounded by a mixture of open parkland and woods. Near the rectory lies the tiny church at Came where William Barnes was rector for over twenty years, during the second half of the nineteenth century. He was also buried in the churchyard here.

Perhaps one of the most famous villages in Dorset is Tolpuddle, due to its historical link with the trade union movement. The village is situated about seven miles to the east of Dorchester. Here, in 1831, six farm labourers collectively attempted to obtain an increase in wages, for which action they were later arrested and transported to Australia. They eventually became known as 'the Tolpuddle Martyrs'. There are many thatched properties in the village and also a thatched shelter with a memorial seat, near the old sycamore tree under which the Martyrs originally met. Many of the cottages are thatched with long straw. A picturesque thatched 'Pixie Cottage' serves teas in the village, just down the road from the Martyrs Inn.

An exquisite collection of thatched cottages can be seen at Milton Abbas, about five miles north of Tolpuddle. This is thought to be one of the first examples in the United Kingdom of an integrally planned village. It was built in the eighteenth century and consists of a single wide street of nearly identical thatched cottages, all uniformly laid out and separated in pairs from one another. It is a most impressive sight when first viewed, and one gets the impression of being transported back in time to another century. A thatched inn, called The Hambro Arms, in the middle of the village, is a long thatched building which conflicts with the apparently uniform series of squarely-built thatched cottages. The village rests in idyllic surroundings in a wooded valley, and about a mile away are historic Milton Abbey and the famous boys' public school. The original village of Milton Abbas was located close to the Abbey Church, and it then consisted of a large number of streets, inns and a brewery. When Joseph Damer, the first Earl of Dorchester, inherited his estates at Milton, he disliked the presence of the village so close to the mansion he proposed to build near the Abbey. He therefore ordered the complete destruction of the old village, and the present-day model thatched village was constructed in its place, a mile away from the site of the old. The former Damer mansion is today the public school. The thatched house found near the Abbey and school constitutes the only original village building which escaped the Damer demolition order.

Many lovely collections of thatched cottages can be viewed in the hamlets and villages nestling in the Piddle Valley, which is located to the west of Milton Abbas and therefore to the north-east of Dorchester. Many of the villages start their names with the word 'Piddle', such as Piddlehinton and Piddletrenthide. (Incidentally, the former has an old thatched country inn called The Thimble.) The use of the name 'Piddle' is derived from the presence of the River Piddle which runs through the valley. The village of Puddletown was originally called Piddletown. The name was changed as it was thought the new one sounded more proprietous. In addition to these villages, other thatched buildings can be found in the hamlet of Plush, a little further north in the Piddle Valley. A long stretch of thatch covers the sixteenth-century inn, The Brace of Pheasants. In keeping with the name, two perched birds sit on the top of the thatched roof which, at eaves level, is attractively cut around the many upper storey windows. The inn was constructed by the alteration and restoration of two cottages and a forge. Alton Pancras rests close by in the Piddle Valley, and a thatched village stores and post office is located in the main street. The walls of the building are constructed of brick and flint; the thatched roof is gabled.

The village of Buckland Newton lies about a mile and a half to the north of Alton Pancras. Many attractive thatched cottages can be viewed in the village, and the occasional thatched house is also fitted with a thatched garage. About half way between Buckland Newton and Dorchester, Dorset claims the smallest public house in England possessing a thatched roof. This can be seen in Godmanstone. The fifteenth-century inn called The Smith's Arms was previously a blacksmith's shop, and King Charles II once stopped there to have his horse shod. He is reputed to have asked for a drink, but the blacksmith could not serve him because he had no licence, so there and then the King granted him one. When viewed from the front, the small thatched roof appears to be conical in shape. The thatch material is combed wheat reed, and the walls of the inn are constructed of cob, faced with flint stones. The front of the inn measures just ten feet across, and the eaves are only four feet from the ground. There are other thatched cottages in Godmanstone, and one house is roofed with an artificial material, which was obviously designed to give the house the superficial appearance of a thatched roof.

The small village of Woodsford is situated about four miles to the east of Dorchester. Here there is an interesting thatched fortified house called Woodsford Castle. The main part of the house was probably built in the fourteenth century, and it is believed to be one of the largest thatched buildings in England. It can additionally claim to be the only thatched castle in England and also the oldest fully-occupied castle. West Stafford lies about three miles away to the south-west and displays several old thatched roofs. One covers the local inn, The Wise Man. Next to the inn, many of the traditional Dorset country crafts can be seen in the thatched Barton Barn Craft Centre. This restored, nearly four-hundred-year-old barn, with a substantial thatched roof, offers for sale many attractive craft souvenirs of interest to the tourist. Items such as corn dollies, pottery, metal work, Dorset feather stitchery and Dorset buttons are typical of these. The thatched building was originally the main barn of the village manor farm estate. The underside of the main roof can be viewed from within the Craft Centre. A series of giant beams and crudely-cut wooden poles support the massive area of thatch. A hipped thatched porchway abuts from the main body of the barn.

A thatched barn of a different type can be viewed at Broadmayne which is fairly close by, to the south-east of Dorchester. This barn is a good example of an old tithe barn. Broadmayne also possesses a sprinkling of thatched houses. A further example of a very large tithe barn can be seen at Abbotsbury, near Weymouth. Unfortunately, although a substantial area of thatch still remains, it is now partly without a roof on one section. It dates from the fifteenth century and is approximately 173 feet long and 30 feet broad.

The vastness of the thatched barn gives a reminder of the former economic system in which payment was made in kind, or large volumes of produce, rather than cash. The village of Abbotsbury consists predominantly of one long street, lined with thatched cottages mingling with a few slate-roofed ones. The walls of many of the cottages are constructed of orange-tinged stones. Many of the roofs are thatched with the local Abbotsbury water reed. In keeping with the long straggling nature of the village, a restored thatched medieval long house forms part of the long terrace. Further along the street, there is a thatched pottery, a thatched antiques shop and a thatched general store. Abbotsbury also boasts some beautiful sub-tropical gardens complete with peacocks and many rare plants. The famous Abbotsbury swannery is also located near the tithe barn, and this, the largest swannery in England, has been in existence from at least the end of the fourteenth century.

The village of Little Bredy hides in a wooded valley four miles north of Abbotsbury, and it is therefore also fairly close to Dorchester. Many delightful cottages, thatched with combed wheat reed, abound in the village. A thatched roof covers the village hall, which also has a separate thatched lean-to roof over a small attached extension. The village hall was at one time the school. Another thatched house occupies the area between the village hall and church. Beautiful grounds adjoin the church, and the public are welcome to walk through them. The church was considerably rebuilt in 1850 and possesses a spire. The stone used in the construction was imported from Caen in Normandy. Two original bells are still in the church: the elder dates back to 1400. A little further north, the village of Frampton is located five miles to the north west of Dorchester. The long village street is made attractive by the miscellany of thatched cottages which lines one side of it. There are no buildings on the other side of the street, which is a distinctive feature. In addition to the thatched cottages, there is also a thatched guest-house in the village.

There are still a few ancient refectory barns surviving in England, where the monks of old used to dine under a thatched roof. One good example can be seen at Toller

Fratrum, about three miles to the north-west of Frampton. This thatched building was once used by the Knights of St John of Jerusalem, the Hospitallers. This was a religious military order which arose out of the crusades to the Holy Land and included Knights, Clergy and Brothers of the Order. The well-preserved thatched refectory barn stands close to a little medieval church. One or two thatched cottages are also to be seen in this rather remote hamlet of Dorset.

The village of Rampisham lies a few miles to the north-west, and the picturesque post office, opposite the inn, is topped with a thatched roof. The many thatched patches on the roof denote its long history of repairs. A small separate canopy of thatch shelters the entrance to the post office. Several thatched cottages and farm buildings border the immediate vicinity of Rampisham. Corscombe is situated a little further to the north-west, and thatched buildings are again common in the area. The Fox Inn at Corscombe represents one of the most impeccable. The inn dates from about 1600. A continuous stretch of immaculate thatch covers the long building, and two small thatched porch canopies protect the entrances. The inside of each bar is tastefully ornamented with a fine collection of country utensils, which include many made of old brass and copper. There are also agricultural implements and prints depicting hunting scenes. (Incidentally, the inn was featured in the film *Rogue Male*, starring Peter O'Toole.)

In the extreme west of Dorset, a very beautiful but peculiar thatched roof can be seen covering the Umbrella Cottage at Lyme Regis on the coast. The cottage was originally built as a toll house. As the name suggests, the thatched roof is shaped like an umbrella. It therefore forms a continuous curve, and there are no flat areas. The eaves are cut in the shape of the curves and points, as they are seen arranged around an umbrella's edge. Two other thatched houses are also prominent on the sea front. One of these, with pink-washed walls, is a little unusual because rainwater collection-guttering is fitted under the eaves of the thatch. Lyme Regis was once a favourite resort of Jane Austen. Mary Anning was also a well-known former resident, who collected many of the fossils that abound in the area; some of the largest and best are now exhibited in the Museum of Natural History in London.

The seaside resort of Charmouth neighbours Lyme Regis. Many thatched buildings stand in the main street, and these include cottages, restaurants and a hotel. Some of the houses exhibit bow windows. The thatched hotel stands at the top of the main street, but also of interest is The Queen Armes Hotel which boasts many historical connections. Catherine of Aragon stayed in the building on her arrival in England in 1501, and during 1651 King Charles II stopped at the inn during his flight from Worcester. The main street runs up a hillside, and the overall picture is most charming. The beach at Charmouth, like that at Lyme Regis, is an excellent place to hunt fossils.

A little further to the east, the village of Burton Bradstock is close to the sea, cliffs and Chesil Beach. This delightful spot contains many stone-built thatched cottages and an inn with a well-kept thatched roof. Chideock is situated just a few miles away, and it also has a miscellany of thatched buildings. In addition to thatched cottages, the main street claims a hotel, a seventeenth-century inn, a restaurant, guest-houses and a mixture of farms, all topped with thatched roofs. The village of Symondsbury, which nestles nearby, off the Chideock to Bridport road, is locally well known for its excellent thatched roofs, and many good examples of sandstone-constructed thatched cottages can be studied there. This peaceful, picturesque village hides in a wooded setting, and it also has a six-hundred-year-old inn, called The Ilchester Arms.

Another six-hundred-year-old inn exists at Shave Cross, a little further north, in the Marshwood Vale. This inn, built with cob walls and a thatched roof, claims to possess the most ancient skittle-alley in Dorset. A thatched porch protects the doorway of the inn. The north-east road from Shave Cross leads to the pretty village of Stoke Abbott, near Beaminster. The distance is approximately three miles, and the approach road to Stoke Abbott is rather tortuous. The main street of the village is very narrow, and almost all the buildings along it are thatched. A thatched post office stands at the end of a thatched terrace row of cottages. A thatched inn has a small extension abutting at right angles to it. However, the extension has a tiled roof in contrast to the main roof thatch. Stoke Abbott also has a twelfth-century church.

A particular roof which would keep a thatcher busy for a very long time, if he had to renew it, exists at the hamlet of Marshwood, near Wimborne, in the east of Dorset. It is perhaps the longest stretch of thatch in England, measuring 360 feet in length. It covers eleven cottages as well as a post office and shop, all joined together in a long terrace row. The walls of the buildings are of cob, although small extensions made of other material have been added. The hamlet of Almer borders the western road from Wimborne. A lovely sixteenth-century thatched inn can be found here, called The Worlds End. The long continuous thatched roof has an ornamental raised ridge. A separate section of thatch covers a right-angle-shaped extension at the end of the inn. The odd name, The Worlds End, is due to the fact that four parishes unite at Almer. Inside the inn, an array of old agricultural implements decorates the bars.

A little to the north-west, on the other side of Blandford, is the pretty hamlet of Hammoon. This contains a good example of a sixteenth-century manor house, furnished with a thatched roof. A short distance to the west lies the large village of Sturminster Newton. This village claims a mixture of thatched houses, including cob and timber-traced constructions. The White Hart Inn possesses a thatched roof. In addition to the thatched roofs, the village offers much of interest. William Barnes, the dialect poet, was born on a farm just outside the village and attended school near the church. Another item of note is the fine medieval six-arched bridge which spans the River Stour and connects Sturminster with Newton. There is also a seventeenth-century town mill which is still in working order. A thatched house stands opposite the mill.

Lulworth Cove attracts many visitors, and it is convenient to reach from both Weymouth and Swanage. The pretty village of East Lulworth has many long straw thatched cottages and houses. Some of the houses are constructed in the shape of a right angle, and the hipped thatched roofs sweep round to cover both sections. The thatched Castle Inn is in the village, and nearby can be found the remains of Lulworth Castle, which was constructed in the sixteenth century but gutted by fire earlier in this century. The hamlet of Kimmeridge borders Lulworth, and most of its cottages are built of stone and topped with thatched roofs. Bituminous shale, found in the cliffs at Kimmeridge, was used by the Romans to make coins and jewellery. Today, an oil rig pumps crude oil from the underlying strata of Kimmeridge. The famous ruins of Corfe Castle tower over the nearby village of Corfe, which has a mixture of stone and thatched dwellings. Some of the thatched roofs have lovely raised ridges. The large holiday resort of Bournemouth is located on the eastern coast of Dorset. Many visitors have viewed the unusual thatched aviary which stands in the main gardens, just off The Square in the centre of the town. Fortunately, the birds in the aviary cannot attack the underside of the thatched roof due to the use of a protective cage top.

WILTSHIRE

Many thatched villages and hamlets surround the city of Salisbury. Typical of these is the village of Steeple Langford, resting in the Wylye Valley to the north-west of Salisbury. This contains several cottages constructed with brick and topped with thatched roofs. In contrast, houses built with flint are also to be seen, and these are designed with a chequered pattern which is a fairly common feature in Wiltshire. Several man-made lakes are located at Steeple Langford, formed by flooding previously excavated gravel pits. The village of Stockton also nestles close by in the Wylye Valley. This picturesque spot has many thatched cottages. It also has several other old buildings, including an Elizabethan flint-constructed farmhouse with a great barn. Some quaint seventeenth-century almshouses may also be seen in the village. To obtain an outstanding view of the Wylye Valley, it is worth visiting Stapleford. This village offers some charming thatched cottages to complement the scenic views.

Teffont Magna is situated to the west of Salisbury and has

several pretty thatched cottages around the church, which possesses a portion of an Anglo-Saxon cross dating back to the ninth century. There are several small bridges crossing the stream in the village, which hides in the wooded valley of the Nadder. In fact, most of the cottages by the stream require individual stone slab bridges to reach their doors.

West Dean lies to the east of Salisbury, overlooking the Hampshire border. Dean Hill rises five hundred feet above the village and offers magnificent views of the surrounding countryside. A sprinkling of thatched roofs can be seen in West Dean, which has a pleasant green and river to complete the scene. Durnford rests in a valley by the River Avon to the north of Salisbury, especially peaceful because there are no main roads to disturb it. Several roofs thatched in long straw may therefore be safely studied at leisure in the village. It will be noted that most of the thatched cottages have walls constructed of flint, and they date to the seventeenth century. The hamlet of Netton is close to Durnford, about a mile along the banks of the Avon. Several good examples of thatched walls and boundary-markers can be viewed in Netton and around its immediate vicinity. Wilsford also nestles close to Durnford, along the Avon Valley. The attractive village displays many timber-framed cottages with thatched roofs. In addition, there are also houses constructed in the more familiar Wiltshire style of patterned stone and flint.

Superb views of the south, towards Dorset, can be gained from the village of Zeals which contains many delightful thatched cottages. The village is to the west of Salisbury, near the borders of both Dorset and Somerset. At the other extremity of the county, a slightly similar opportunity is afforded at Liddington, just to the south-east of Swindon. The village has a large number of pretty thatched cottages and from Liddington Castle wonderful views of three different counties, Oxford and Berkshire as well as Wiltshire. In addition to the thatched cottages, Liddington also possesses a seventeenth-century manor house, which has an attractive pond created from a section of the original moat. In contrast to these border villages with views over several counties, the village of Tilshead is set in the open spaces of Salisbury Plain.

It contains many stone and flint thatched cottages; the walls are chiefly constructed in the chequered-patterned style. A few of the cottages boast ornamental pheasants on their roof ridges. Many ancient barrows and earthworks are gathered in this area of the Salisbury Plain, to the north of Stonehenge. A very long excavation, called 'The Old Ditch', can be found near Tilshead. There is also the White Barrow, approximately three hundred feet long, to the south of the village. These barrows were once used as Neolithic burial places, and the one at Tilshead is probably the longest barrow in England.

The village of Horningsham stands to the south west of Warminster. A small thatched chapel called The Old Meeting House is still in use here. The chapel dates to the second half of the sixteenth century. It is interesting that the thatched chapel was originally built by Scottish workmen who had come down to Wiltshire to construct the 'stately home' of Longleat House, which is located close by. The Old Meeting House is perhaps the earliest Presbyterian chapel in England. The date 1566 is distinctly marked on the end outer-wall of the chapel, just below the eaves level of the thatch and centrally above the large end windows. The thatched roof has been kept in immaculate condition and is capped with a neat raised ridge possessing points. At eaves level the beautiful thatch sweeps in a smooth contour over the upper windows of the chapel.

In the northern part of Wiltshire, an old malthouse with a thatched and partially tiled roof can be seen at Aldbourne, near Marlborough. The roof of the kiln area is fitted with tiles, and this was no doubt originally done on safety grounds with regard to fire risk. The village, although large, is one of the most lovely to be found in Wiltshire. Several thatched cottages with colour-washed walls stand near the village green. Also nearby are an ancient stone cross, a pond and the church, which was built in the twelfth century. An unexpected sight can be discovered inside the church: two eighteenth-century fire engines. The village was quite famous once for straw and willow plaiting. It was also well known for its bellfoundry, mainly making small bells for horses and farm animals.

Avebury is situated to the west of Marlborough and is renowned for the huge Avebury Stone Circle which surrounds it. Unlike those of Stonehenge, the stones at Avebury show no sign of having been worked. The circle was built at an earlier time than that at Stonehenge, probably during the late Neolithic period, around 2000 BC. The village of Avebury has many thatched cottages and a thatched great barn. It also possesses an Elizabethan manor house.

The towns of Pewsey and Bedwyn stand by the side of the canal and railway to the south of Marlborough. Both retain thatched properties, and Pewsey still has some shops which are thatched. It also has a rare example of a medieval cruck house with a thatched roof. The house is of half-timbered construction and is winged with cruck-ended gables. Long straw constitutes the thatch material used for the roof. Many thatched cottages, houses and farm buildings also lie between Pewsey and Bedwyn. Several of these are located very close to the railway line. In the days of the steam engine, they must have been subjected to some fire risk from the sparks emitted from passing trains. However, most appear to have survived, which illustrates that thatch is perhaps more difficult to ignite than one might expect. The thatch in this region is mainly long straw, although some reed is also evident.

Castle Combe, situated a few miles to the north-west of Chippenham, gives the impression to its many visitors that it must be the most beautiful village in England. It is located in a wooded valley, and a little stream meanders under the arched bridge. Gathered nearby are an ancient market cross and a picturesque collection of cottages grouped around the church. The walls of the cottages are constructed of Cotswold stone; some are roofed with thatch and others with stone tiles. The main hotel in the village was originally a seventeenth-century manor house. (The village of Castle Combe was used on location during the making of the film *Dr Doolittle.*) The ancient Roman Fosse Way passes by the western side of the village.

Lacock, three miles south of Chippenham, rivals Castle Combe for consideration as one of the most beautiful villages in England. All the houses in Lacock are old, and they span many centuries of English architectural styles, from the medieval period to the eighteenth century. Some of the houses are stone built, others are timbered buildings, and several possess thatched roofs. A fourteenth-century barn overlooks the hotel. Lacock Abbey belongs to the National Trust, together with the rest of the village. The abbey dates to the thirteenth century but was converted into a house during the sixteenth century.

Calne sits between Chippenham and Marlborough. Many people know the name of the town because of its thriving, world-famous bacon industry. A little south-west of the town can be found the village of Sandy Lane. The nineteenth-century Church of St Nicholas in the village has a thatched roof to protect its timber-framed construction. A thatched porchway guards the entrance to the church. In addition to the church, there are many thatched cottages in this delightful spot. The unusually named Clyffe Pypard perches on a tree-covered ridge to the north-east of Calne. This village also boasts many thatched cottages.

AVON

A neatly-built collection of thatched cottages and other buildings can be viewed on the Blaise Castle Estate, which is sited a few miles north of Bristol. In this four-hundred acre woodland estate there can be seen an early nineteenth-century dairy, which has been retained in its original thatched form. There is also a hamlet of nine thatched cottages gathered around a village green and pump. They were also built during the early nineteenth century, when thatch was fashionable for the construction of planned, integrated model villages throughout England. Other buildings on the estate include a castle with four battlemented towers, erected as a folly in the late eighteenth century and after which the estate was named. There is a corn mill on the estate and a Folk Museum, which exhibits, among other items, a typical old farmhouse kitchen.

Portishead, to the west of Bristol, overlooks the sea from a tree-covered hillside. One of the oldest buildings in this popular resort has a thatched roof and also the remnants of a moat around it. The thatched building is known as The

Grange and can be found in the High Street. It was once lived in by the manor bailiff.

A thatched roofed former toll-house guards the road junction leading to Stanton Drew, from the Chew Magna road to the south of Bristol. The thatched house demands attention because of its unusual hexagonal shape. On the eastern outskirts of Stanton Drew, a series of ancient stone circles stand in a field. The three circles consist of massive upright stones which are thought to date from approximately the same period as those at Avebury. Also of interest at Stanton Drew is the medieval rectory with Gothic windows situated near the bridge over the River Chew.

Badminton House stands in the village of Badminton, five miles east of Chipping Sodbury. The Palladian-style house has been the home of the Dukes of Beaufort since the seventeenth century. It is open to the public on many days of the year. The park grounds are internationally famous as the venue for the Three Day Event Horse Trials, often attended by the Queen. The hamlet of Little Badminton borders the edge of the park. A fine collection of tiny old thatched cottages surround the green. It is believed that many of these date back to at least the sixteenth century. An ancient turreted dovecot dominates the green. It is said to possess the same number of nesting holes as there are days in the year.

HAMPSHIRE

Combed wheat reed remains a popular thatch material in Hampshire, especially in the western regions, but long-straw thatched roofs are very frequently encountered, and local Hampshire marsh reed and Norfolk reed are also utilized to a more limited extent.

Many thatched villages surround the town of Fordingbridge, located in the extreme west of Hampshire. Some of the communities to the north-west of Fordingbridge formerly belonged to Wiltshire, and typical of these is the charming village of Martin, where much thatch and cob may be seen. About three miles away to the east can be found the tiny village of Rockbourne. A stream meanders through the main street which is bordered by several thatched cottages of brick and timber-framed construction. The manor house in the village is partly Elizabethan, but it has a fourteenth-century barn. The remains of an excavated Roman villa stand about half a mile away. This originally consisted of a very large number of rooms, baths and mosaic pavements. Another three miles to the east, on the remote edges of the New Forest, may be discovered one of the forest region's most beautiful villages. This is Breamore, which has several thatched Tudor cottages positioned at the corner of the village green. Some of the roofs are thatched with reed, others with long straw. Thatched porches are also to be seen. Incidentally, one of the few remaining Anglo-Saxon churches in Hampshire still stands in the village. The village of Fritham is located to the east of Fordingbridge. A small thatched inn, called The Royal Oak, can offer hospitality. Approximately one mile outside the village, an unusual collection of huge ancient yew trees may be seen at a place called Sloden Enclosure.

South of Fordingbridge, the Ringwood road passes through Ibsley. The thatched cottages in this village are ancient enough to have been mentioned in the Domesday Book. A bridge and several weirs provide an attractive setting for the village. Nearby, on the River Avon, Ringwood has some pretty thatched cottages, and the houses in the town date from many different periods of English history. A picturesque thatched licensed restaurant, called The Old Cottage, can be found in West Street. A pleasant row of thatched cottages lines the road by the bridge. Also near the bridge can be seen a property named Monmouth's House. This house became the temporary residence of the Duke of Monmouth during his abortive rebellion and immediately after his capture. He was later taken to London and executed on the block. A pleasing thatched inn, called The Sir John Barleycorn, welcomes visitors just outside Ringwood in the New Forest. The property is covered with a long sweep of thatch, and three individual thatched porches guard the entrances to the inn.

One of the oldest inns in England may be found at Hinton Admiral, which is located to the south of Ringwood. The inn, called The Cat and Fiddle, is over six hundred years old. It is constructed with cob walls and roofed with thatch. Another old thatched inn, The Fleur de lys, exists at Pilley, sheltering a little further east in the New Forest. There was much smuggling at one time in its immediate area. Nearby is Brockenhurst, which makes a good centre for touring the New Forest. This small town also offers some pretty thatched cottages. In addition, it has a good example of a church which was built on a mound, once quite a common practice in the surrounding parts. The churchyard possesses a giant yew tree which is thought to be nearly a thousand years old. Minstead, near Lyndhurst, has a beautiful secluded woodland setting. Sir Arthur Conan Doyle, the creator of Sherlock Holmes, lived in the village and is buried in the churchyard. Furzey Gardens, just outside Minstead, displays a magnificent collection of shrubs and flowers. A quaint folk-cottage, built in 1560, stands in the gardens, and a long-straw thatched roof caps it.

Many thatched villages abound in the region of Stockbridge and in particular along the banks of the River Test, which flows through the Test Valley. The river constitutes one of England's most superior waters for trout fishing. At Longstock, just a mile and a half to the north of Stockbridge, some picturesque thatched huts line the river bank. They are used by anglers for storing their tackle. The village of Longstock itself consists of one long street of houses, many of which are thatched. The walls of the houses are constructed of a contrasting mixture of materials. Many are colour washed but there are also timber-framed and red brick houses all in close proximity to one another. The village of Leckford nestles about a mile away in the Test Valley. This contains some thatched timber-framed cottages, several of which have elegant raised ridges with points. Thatched timber-framed cottages can also be found in the charming village of Wherwell, which is about three miles to the north of Leckford. After a visit to Wherwell, it will be difficult to dispute the claim that it is the most lovely village in Hampshire. It displays many old impeccably thatched timbered cottages, set in a wooded background. An exceptionally good view of the Test Valley may be obtained from the west side of the village.

The small village of King's Somborne rests to the south of Stockbridge, along a branch of the River Test. An old thatched inn, called The Crown, borders one side of the village green. The church, dominating the other side, possesses two rare fourteenth-century brasses depicting men wearing civilian cloaks and carrying short swords. The brasses are exceptional because the figures are not dressed in priest's garments or armour, as is the normal fashion. Nether Wallop is situated to the west of Stockbridge, and other typical examples of timber-framed thatched cottages may be seen here. There is also a mill in the village. About two miles away is the village of Broughton which has many timbered houses and several old farmsteads which have their boundary walls capped with thatch.

A thatched wall may also be seen at Sutton Scotney, through which runs the main north road from Winchester. The wall adjoins the garden entrance to the Sutton Manor Estate, and it is thatched with marsh reed. The little village of Crawley is situated fairly close by, to the north-west of Winchester. Again thatched timber-framed cottages can be viewed along the single main street, and these are mingled with other houses built at different periods. The village of Easton, on the east side of Winchester, possesses many thatched dwellings and also a church which dates from approximately the end of the twelfth century. The village could be visited on the way to Tichborne which rests a little further east of Winchester. Tichborne is a charming, unspoilt spot containing many thatched properties and sixteenth- and seventeenth-century houses.

A magnificent example of a thatched tithe barn can be studied at Hensting Lane, near Owslebury on the south-east of Winchester. Three tall, heavy buttresses support the end wall of the old barn, which has a Sussex hip type of thatched roof over it. The main thatched roof, covering the long sides of the barn, sweeps down low to about six feet above ground level. In West Meon, further to the south-east of Winchester, are some beautiful thatched gabled cottages constructed with plastered and timbered walls. The River Meon runs by the village, and it may be of interest to cricket lovers that Lord, after whom the famous cricket ground in London is named, is buried at West Meon.

THE ISLE OF WIGHT

Across the water from Hampshire, on the Isle of Wight, the main seaside town of Shanklin is skirted by the Old Village. This is a picturesque village which contains many thatched cottages with roses and honeysuckle climbing their white-fronted walls. There is also a thatched restaurant on the main street. A beautiful wooded chine links the Old Village with Shanklin. A further village which attracts many visitors is Alverstone Mills, about three miles from Sandown. There are several thatched stone-built cottages in the village, and many offer garden teas during the summer months. There is also an old water mill, and boating is available on the stream. Another village full of thatched cottages and flower gardens can be found at Newchurch which is located about two miles further west.

The main road between Sandown and Newport passes through Arreton, about three miles to the west of Newchurch. Arreton has the old thatched inn, The Hare and Hounds, to dispense hospitality. The farm attached to the manor house has an old large thatched barn. Godshill is situated to the south of Arreton; its many tidy tea-gardens and quaint old thatched cottages attract hosts of summer holiday visitors. Most of the cottages are thatched with long straw. An exceptionally photogenic group stands near the entrance to the churchyard. The church contains a very rare wall-painting, known as 'the Lily Cross', dating from approximately 1450. Other items of interest in the village include a model village situated in the Old Rectory garden and an exhibition of shells. A further four miles to the north west, the village of Gatcombe retains its unspoilt character. Stone-walled thatched cottages are to be seen, nestling among the trees and near the stream. A magnificent Palladian-style house, set in parkland, overlooks the village.

Towards the western side of the island, the village of Calbourne provides excellent picnicking areas in its immediate vicinity, north of the Brighstone Forest. A quaint old-world atmosphere prevails through Winkle Street in the village. The street includes some low stone-constructed thatched cottages, and flowers bedeck the walls of these and the adjoining houses. A little stream flows by to complete a much-photographed picture. About three miles to the south, the village of Brighstone rests a few miles inland from the waters of Chilton Chine, which offer relatively safe bathing at high tide. The village of Brighstone has many tea-gardens and old stone-walled thatched cottages. A valley to the east hides the village of Shorwell. This also contains many picturesque old stone-walled thatched cottages, mainly gathered by the inn near the stream. There are some Elizabethan manor houses in the neighbourhood, which is rather strange for such a relatively small village.

Freshwater Bay, on the more extreme west side of the island, claims a thatched church. This was built at Freshwater Gate at the beginning of the twentieth century; it is therefore relatively modern for a thatched church. The land on which it was built had former associations with Tennyson, the poet, who lived in the village of Freshwater for many years.

BERKSHIRE

The village of Pangbourne nestles along the banks of the River Thames at the junction where the Pang, a small trout stream, meets the main river. The village, situated to the west of Reading, remains attractive despite the encroachment of modern building development. Several seventeenth- and eighteenth-century houses, together with some lovely thatched cottages, still survive in a pleasant riverside setting. The Pangbourne Nautical College sits on the top of the hill. Good views of the countryside and river may be obtained at the weir and river lock, but for really spectacular views of the countryside, it is better to travel a little further north-westward to the village of Streatley. This Thames riverside village shelters in a deep gap which separates the Berkshire Downs from the Chiltern Hills and yields a good vantage point from the high ground at the back of the village. Some thatched properties are to be seen at Streatley, together with Georgian-built houses and a malthouse constructed in the nineteenth century. A bridge crosses the Thames at Streatley and leads to Goring in Oxfordshire.

In the opposite direction, the west road from Streatley rises to the peaceful village of Aldworth. This presents thatched cottages and a church which was once visited by Queen Elizabeth I. A yew tree has stood in the churchyard for about one thousand years, but it is now in an advanced state of decay. Aldworth also possesses a canopied well which descends 372 feet below the surface, to make it one of the deepest in England. Sonning rests on the banks of the Thames but to the east of Reading. The village boasts many delightful timber-framed cottages, topped with thatched roofs. An old mill also still survives in the village. The bridge which crosses the river at Sonning, to Oxfordshire, is one of the oldest spanning the Thames.

The village of Leverton borders the northern outskirts of Hungerford in the west of Berkshire. A row of nearly identical thatched-roofed cottages may be viewed at Leverton. They were built at the beginning of the nineteenth century, along the lines of a planned integrated village. The thatched roofs conflict with the slate roofs over the porches of the cottages. The use of the two materials may have been a fashion compromise. At the time the cottages were built, thatch was becoming fashionable again as slates were in plentiful supply throughout England.

A short distance to the north, the Lambourn Downs provide excellent gallops for the training of race-horses. The limestone below the turf ensures a firm and good draining strata to maintain good going for horses. The River Lambourn runs through the valley, and many pretty thatched cottages are sprinkled along its length. The hamlet of Eastbury rests two miles downstream from Lambourn. Several thatched timber-framed cottages border the river, reached by quaint little bridges spanning the stream. The village of East Garston is situated a little further downstream and displays several thatched cottages along its twisting street. Welford rests a little further south in the Lambourn valley and also has many attractive thatched cottages.

The Berkshire Downs are located to the north-east of Lambourn and are equally famous for the training of race-horses. Again many thatched cottages are scattered throughout the vicinity. The hamlet of Beedon, to the south of East Ilsley, also provides a row of old thatched timber barns which are raised on brick bases. They are found near the church, which has a huge fourteenth-century open-timber roof over the nave.

GREATER LONDON AND SURREY

The number of thatched buildings in the London area is very small, but the occasional one may be discovered in parks or public gardens. The Royal Botanic Gardens at Kew offer Queen Charlotte's thatched cottage, which is open to the public on certain times during the summer months. The thatched roof covers the two-storey cottage, constructed of brick with timber facing boards. The cottage dates from 1772 and was built in the rustic style which was popular with the wealthy at that time. It was used by Queen Charlotte as a summer house to give tea parties and picnics. A further royal connection with the name of thatch exists at Richmond Park, where The Thatched Cottage was the former residence of Princess Alexandra. Bushey Park, which adjoins the gardens of Hampton Court Palace, also claims a thatched dwelling. This is situated along the line of the perimeter wall on the Teddington side of the park in Hampton Wick. The long thatched building runs parallel with the road bordering the park wall.

Claremont House stands about half a mile away from Esher, along the Oxshott road. This Palladian-style house was built in 1772 for Clive of India. Its fine gardens were laid out by 'Capability' Brown. The house and grounds are open to the public on limited occasions, as it is now used as a Christian Scientists School for girls. Claremont also names a neighbouring private estate which contains some beautiful detached thatched homes. This exclusive estate faces its own private golf course and also has the benefit of the nearby Oxshott Woods.

SUSSEX

The Weald and Downland Open Air Museum spreads over an area of nearly forty acres, located at Singleton, approximately five miles north of Chichester in West Sussex. It exhibits a collection of historic buildings from the south-east of England, which have been saved by the museum from total destruction and re-erected. These include a fifteenth-century Wealden Hall, a fourteenth-century farmhouse, an eighteenth-century granary and a nineteenth-century toll cottage. Other exhibits include a sixteenth-century tread-wheel and a reconstruction of a Saxon weaver's hut, which illustrates the early all-roof-and-no-wall-type construction. The roof is thatched, and the eaves reach ground level. The museum provides picnicking areas and nature trails through its wooded parkland.

The Selsey peninsula lies to the south of Chichester. Water surrounds Selsey on three sides, but despite its rather exposed position, thatched properties can be found on the peninsula. The small Thatched House Hotel, built on the sea front, has its own garden which leads to the sea. North of the Selsey peninsula, some picturesque thatched houses sprinkle the village of Sidlesham. Just a few miles along the coast, Aldwick nestles close to the western suburbs of Bognor Regis. Several old thatched cottages stand in their well-kept gardens to make a delightful scene. A good example of a modern Gothic church is also to be found at Aldwick. Felpham adjoins Bognor Regis on its eastern suburbs, but it has managed to retain its individual character. Several thatched-roof cottages still enhance the village street. A rather famous thatched white-walled cottage stands in Blake's Road. As the name suggests, William Blake, the poet and painter, lived there for some years (at the beginning of the nineteenth century). A little further eastward, the village of Rustington neighbours the coastal resort of Littlehampton. This also boasts a fair collection of old thatched properties. The nearby village of Poling has many thatched brick and flint cottages.

The hamlet of Bignor is about five miles to the north of Arundel. A fifteenth-century yeoman's house, roofed with long straw thatch, may be viewed near the church. The house during its later history became a shop, but it is now a private home. The walls supporting the thatched roof are of timber-framed construction. The wood is oak and the in-filling bricks, although some flints are also present. They are built on an unusually high stone foundation, which necessitated the building of steps to reach the front door of the house. One of the largest Roman villas built in Britain was discovered near Bignor, and the site, covering several acres, draws many visitors. The excavated villa shows fine examples of mosaics, and a small museum, constructed from one of the original rooms, exhibits a miscellany of items found on the site. The village of Amberley is located about six miles to the east of Bignor. Several old thatched brick and flint cottages line the twisting lanes of this lovely village overlooking the River Arun. Amberley also has the remains of a Norman castle and a towered gatehouse which was built during the fourteenth century. A little to the north, a historic thatched cottage exists at the village of Fittleworth. This is 'Brinkwells', where Elgar lived and composed several of his major musical works during the latter part of the First World War.

An intriguing thatched cottage stands at Henfield, to the north-west of Brighton. It is a sixteenth-century building known as 'The Cat House'. Many metal cats adorn the timbered-framed walls just below the eaves level, and the hipped roof is thatched with long straw. Local legend suggests that the metal cats were originally placed on the walls to remind the vicar that his cat had killed the owner's pet canary. The Cat House stands in the lane leading to the church. There are many other old interesting buildings and inns to be found in Henfield, some impeccably thatched.

Lindfield sits on the outskirts of Haywards Heath, near the border of West and East Sussex. The Thatched Cottage, located next to the church at Lindfield, is a close-studded Wealden house in an excellent state of preservation. It is believed that the cottage was once used as a shooting-lodge by Henry VII. Borde Hill Gardens lie about two miles to the north-west of Lindfield. These beautiful gardens and woodlands are open to the public on certain days of the year. They contain many rare trees and shrubs, magnificent magnolias and camellias. The gardens at Heaselands, situated one and a half miles west of Haywards Heath, are also open to the public on certain limited days. These are undulating water gardens, with waterfowl and an aviary within them.

Many picturesque old thatched cottages beautify the northern part of the village of Rodmell, which is found to the south of Lewes in East Sussex. The village also offers a very lovely church, and a visit inside is rewarding. On a historical note, Rodmell claims to be one of the selected areas in England where mulberry trees were first grown during the beginning of the seventeenth century. The leaves of the mulberry tree were used for feeding silkworms, and Rodmell eventually developed a thriving silk industry. The first house purchased by the National Trust stands at Alfriston, four miles north-east of Seaford. This is the Clergy House, built in the middle of the fourteenth century for the parish priest. A thatched roof still protects the half-timbered and wattle-and-daub house which is located beside the church.

Many old timbered houses remain in the village of Northiam, which is about twelve miles to the north of Hastings. A thatched roof covers Silvenden Manor, built with timber-framed walls. The house is believed to date from the middle of the fifteenth century, although it possesses Tudor-style arches. Another manor house, Great Dixter, is not thatched but has one of the largest and grandest timber-framed halls to be found in England. Yet another historical timbered house exists at Northiam, Brickwall, which, like Great Dixter, is opened to the public. Another item of note at Northiam is an oak tree, named after Queen Elizabeth I, which grows on the village green. The Queen dined beneath its boughs in 1573.

KENT

The number of thatched dwellings in Kent is very small in comparison with Sussex. Tiled hung and pantile roofs constitute a more familiar sight. At one time, Romney Marsh formed a good source of cultivated water reeds for thatching purposes, but the reclamation of land and the development of the Romney Marsh area for other farming purposes has, over a long period of time, replaced the large areas of reed beds. Only relatively small patches of reeds now remain.

Newington borders the western suburbs of Folkestone but still retains its pretty village character. A well-known building in the village, called Frogholt Cottage, has formed the subject of many artists' paintings. The quaint thatched building possesses an overhanging storey which gives it considerable charm. Goodnestone rests in the centre of delightful parkland to the west of Deal. The post office in the village has a thatched roof. The building was formerly used as a rectory and was constructed in Tudor times. It is the oldest building in the village, with the exception of the thirteenth-century church. Just two miles to the north, the village of Wingham is halfway between Canterbury and Sandwich. This charming spot has a thatched restaurant in the High Street, The White Cottage Restaurant, which dates from the sixteenth century. There is a park just outside the door. Much further westward, the village of Borden nearly adjoins the town of Sittingbourne. In the centre of Borden, near the church, a large barn with a thatched roof still remains.

(9)

Thatched Buildings in the South Midlands –

Oxford, Gloucester, Warwick, Hereford and Worcester, Hertford, Buckingham, Bedford, Northampton

OXFORD

The county of Oxford contains a very large number of thatched dwellings, and the majority of these are thatched with long straw. There are many villages in the near vicinity of the city of Oxford which display a profusion of thatched roofs. To the south-east, at the village of Chalgrove, an unusual street of thatched cottages can be seen. A stream runs down each side of the road, and the cottages are reached by a series of tiny bridges. Just outside the village, an obelisk marks the spot where a decisive battle was fought in 1643, during the Civil War. To the south of Oxford, Radley has some houses which are of half-timbered construction and roofed with thatch. The village is positioned along the banks of the River Thames, and on the opposite side of the river, Nuneham Woods stretch along the bank for two miles. These beautiful woods are linked by a rustic bridge to a wooded island, in the middle of the Thames. A thatched cottage rests by the island, and its lawn extends down to the banks of the river. The delightful scene of rustic bridge and thatched cottage makes a favourite subject for artists' brushes and pens. Another pretty bridge, and one of the oldest, spans the Thames to the north of Kingston Bagpuize. Also near to Kingston Bagpuize, which lies south-west of Oxford, is a quaint old thatched toll-house which has a rounded frontage. The village also presents visitors with the opportunity to view the gardens of Kingston House. These are open to the public, on certain occasions during the year, to display bulbs, flowering shrubs, roses and herbaceous borders.

Many thatched villages surround Didcot to the south of Oxford, and Didcot itself retains many charming thatched dwellings. A mile south-east of Didcot can be found East Hagbourne, a picturesque village possessing many timber-framed cottages with thatched roofs. Bricks, arranged in zig-zag fashion, form the wall in-filling on several cottages. Some of the neat gardens of the cottages are enclosed with walls constructed of cob and are capped with thatch. The whole village was rebuilt in the seventeenth century, after the old one, with the exception of the church, had been destroyed by fire. The smaller neighbouring village of West Hagbourne also offers a quaint collection of half-timbered thatched cottages. These are set along the twisting roads and beside the village pond. The village of Blewbury rests two miles to the south, but its peace has been rather shattered in recent years by the main road passing through it. However, the village still retains much of its original character, with many old houses and cottages displaying thatched roofs. Thatched walls may also be seen, bordering the orchards and gardens of the village. Aston Tirrold is set amid beautiful countryside, a little to the east. Both thatch and tiles cover the old cottages which are mostly built of cob, although a few are half-timbered.

Harwell, to the west of Didcot, comprises a combination of the very new and the very old. The Atomic Research Establishment symbolizes the modern-day world, while the many old buildings in the village depict our past. One of the few remaining thatched timber-framed cruck houses in England still survives at Harwell. The thatched roof spreads over the arched timbers which were constructed in the fifteenth century. Other old buildings in Harwell include the church, parts of which date from the twelfth century, a fourteenth-century farm and some eighteenth-century almshouses. A little to the west of Harwell, East Hendred contains some attractive timber-framed houses and cob cottages which are thatched. Long Wittenham, situated to the north-east of Didcot, also displays some houses which are beautifully thatched. Magnificent views of the surrounding countryside may be obtained at the nearby hamlet of Little Wittenham.

About a mile to the north of Long Wittenham, the village of Clifton Hampden sits by the side of the River Thames. This preserves some immaculately thatched homes and also the rather famous thatched Barley Mow Inn. This was featured by J.K. Jerome as one of the settings for his *Three Men in a Boat*. Jerome once stayed in a room of this quaint latticed-windowed inn. A cruck construction supports the thatched roof, which sweeps down to an eaves level fairly close to the ground. There is another elegant thatched hostelry, at Clifton Hampden, The Plough Inn. A delightful sweep of thatch, topped with a raised ridge, covers this old timbered building. An attractive nineteenth-century six-arched brick bridge spans the Thames nearby. A short distance downstream rests the village of Dorchester, and the tiny hamlet of Overy, to which it is linked, can be reached by crossing the bridge. This hamlet has an exquisite setting, and the many old thatched cottages further enhance the scene.

Thomas Hardy based his *Jude the Obscure* on the village of Letcombe Bassett which lies to the south-west of Wantage. Hardy used the name 'Cresscombe' for the village, which is famous for its watercress beds. Arabella's Cottage at Letcombe Bassett is a detached long-straw thatched building and it was here that Hardy's Arabella lived. The neighbouring hamlet of Letcombe Regis also conceals some delightful old thatched houses. The village of Sparsholt is situated three miles to the west of Wantage, and here half-timbered constructed cottages, roofed with thatch, may be seen. The village also possesses an unusually large church, some of which dates back to the twelfth century. A further three miles to the west, the village of Woolstone hides in a dip on the Downs. This picture-book village displays some lovely thatched dwellings.

The village of Cassington lies about six miles to the north-west of Oxford. This is also a pretty village, and a collection of stone-walled thatched cottages are gathered around the village green. Approaching the remote Otmoor area, to the north-east of Oxford, one finds the village of Stanton St John, which has several good examples of thatched farms and barns. In this village was born John White, the founder of the State of Massachusetts. Two miles away, on the eastern edge of Otmoor, the village of Studley has some lovely thatched timber-framed cottages. An Elizabethan house, called Studley Priory, also stands in the village. Another village, Islip, on the border of Otmoor, was where Edward the Confessor was born in 1004, when a Saxon palace existed there. Islip has several picturesque stone-walled thatched cottages.

Further stone-walled thatched dwellings cluster in the village of Great Tew, which is located to the north-west of Oxford, near Chipping Norton. Great Tew, set in a park woodland setting, constitutes one of the loveliest villages to be found in the county of Oxford. Magnificent long-straw thatched roofs, with scalloped ridges, cover many of the seventeenth-century cottages. Others possess stone roofs and mullioned windows. The village stocks, also dating from the seventeenth century, still remain on the green. The hamlet of Sarsden lies three miles to the south-west of Chipping Norton. Some exclusive thatched homes, with beautiful scalloped ridges, may be found built here in their own extensive grounds. Most again date from the seventeenth century. Further south, by the River Windrush, can be found the village of Taynton. Most of the cottages are kept in immaculate condition and some of them are thatched. Sir Christopher Wren visited the local quarries, and the stone used for the building of St Paul's Cathedral came from this region.

In the extreme north of the county, the village of Wroxton is three miles to the north-west of Banbury. Most of the houses there are built with stone, and many are also adorned with mullioned windows. The roofs consist of thatch or stone. Near the village, Wroxton Abbey stands by the main road. This gabled house was built during the early seventeenth century, on the site of an old priory. Inside, an embroidered quilt worked by Mary, Queen of Scots, is exhibited. The house also contains a bedroom in which George IV once slept. A short distance away is the village of Hanwell. This exquisite spot has not only thatched and stone-built cottages but also a fine Tudor castle. The main road, south of Banbury, runs through the village of Adderbury. The magnificent church spire dominates the many thatched cottages and houses in the village. Many of these thatched-roof properties were built during the seventeenth century.

GLOUCESTER

In the county of Gloucester, the number of dwellings roofed with the local Cotswold slates outweighs the number of properties with thatched roofs. However, the pitches of the roofs fitted with Cotswold slates are very steep, just as one would expect with a thatched roof. The high pitch is required because the slates consist basically of split pieces of limestone, and these can prove porous if water is not quickly shed from their surfaces. A typical Cotswold-slated roof has small slates at the ridge area, with larger-sized slates at the eaves. In fact, there is a progressive increase in slate size from the top to the bottom of the roof. There is thus no smooth uniformity over the whole area as is seen with a thatched roof.

Five miles south-west of Cirencester, there are many attractive thatched cottages in the village of Tarlton, most of them built of stone. The countryside around Tarlton is pretty, and several roads bordered by trees meet at the village. Also located near Cirencester, but three miles to the east, is Ampney St Peter. The houses here are constructed of Cotswold stone and most have thatched roofs. Tiny dormer windows peep through the thatch to give the houses a charming and quaint appearance. The church in the village claims parts which date back to the Saxons. About six miles to the north-east of Cheltenham can be found the village of Gretton. This delightful place has several old thatched cottages in the close vicinity of the Old Tower.

Further north in the county, approximately two miles to the east of Chipping Campden, the Cotswold village of Ebrington reclines on the hillside. It contains many thatched cottages set amid a lovely wooded background. Again, most of the cottages are built with stone. Very close by, at Hidcote Bartrim, the National Trust administers Hidcote Manor Gardens, one of the most beautiful English formal gardens. They can be viewed by the public. The small village of Ryton borders the motorway in the north-west corner of the county. In its more peaceful days, the poet Abercrombie lived and worked in one of the small thatched cottages in the village.

WARWICK

The western border areas of the county of Warwick are particularly well endowed with timber-framed buildings. House walls built of sandstone are also encountered throughout the county, together with thatched roofs. There is a great number of thatched villages and hamlets in the vicinity of Stratford-on-Avon, and it is worth observing that the chimneys of the older buildings in the town itself are exceptionally tall. They were originally erected in this way so that sparks emitted from the chimneys could be safely carried away, before falling on their roofs, which were once thatched.

Many tourists visit the most famous thatched building in the county, Anne Hathaway's Cottage at Shottery, snuggling close to Stratford-on-Avon on its western perimeter. Anne Hathaway was born in the thatched cottage and lived there until her marriage to William Shakespeare in 1582. The cottage was previously a fifteenth-century yeoman's farmhouse and is relatively large, containing twelve bedrooms. The thatched roof has a ridge with a raised, pointed bottom edge. It is tastefully decorated with cross rods to give a diamond-shaped pattern. These blend with the latticed windows of the black and white timber-framed house. The village of Shottery also possesses many thatched timber-framed cottages, set in their own individual neat gardens.

About three miles to the south-west can be found the pretty village of Welford-on-Avon. This contains many thatched and timber-framed houses along the main street, sloping down to the river. A seventy-foot-high striped maypole stands on the village green, topped with a weather-vane in the shape of a fox. Three miles downstream, the Avon reaches the village of

Barton, which has some excellent examples of thatched houses; the occasional one displays an ornamental straw weathercock. A short distance further south-west, at Salford Priors, several well-maintained thatched seventeenth-century timber-framed farm houses may be seen.

The village of Lower Quinton can be explored five miles to the south of Stratford-on-Avon. Some charming timber-framed houses, with thatched roofs, surround St Swithin's Church, which has an unusually tall spire. Many people rate Ilmington the most beautiful village in the county of Warwick. It is situated another three miles to the south of Lower Quinton. Visitors enjoy the sight of several stone thatched houses and cottages, a gabled manor house and a church which has many late additions, although it predominantly dates from 1500. Hurdle-makers still work in the village, and this craft has survived as well as that of the thatchers. Another picturesque village lies five miles to the east of Stratford-on-Avon. This is Wellesbourne which now comprises Wellesbourne Mountford and Wellesbourne Hastings. The Stag's Head Inn stands in the village under a thatched roof. The nearby hamlet of Butlers' Marston, to the south-east, is also worth a visit, to see its lovely collection of thatched cottages.

The village of Whichford rests further south, near the county border. Several thatched yeoman farmers' houses line the village green. The houses are mullion windowed and stone walled, and some also possess stone roofs, which give variety to the adjoining thatched ones. Also of note in the village is Whichford House, near the church, which possesses a beautiful English country garden. Long Compton is only a couple of miles away, and the place is of interest because there is a thatched churchroom above the lych gate of the churchyard.

Edge Hill towers seven hundred feet over the peaceful village of Radway in the south-east of the county. Trees surround this village of charming thatched cottages clustered around the green. The gabled Radway Grange in the village was once owned by an ancestor of George Washington, President of the United States of America. In 1642, Edge Hill was the site of a famous Civil War battlefield in which the Royalist army confronted the Parliamentary forces. The village of Tredington, to the south-west, still suffers visible evidence of the Civil War. The huge wooden door of the church bears battle-marks, and the remnants of some bullets are still embedded in it. Opposite the church, which dates to Saxon times, is an attractive long-straw thatched roof.

HEREFORD AND WORCESTER

The county of Hereford and Worcester has a number of thatched dwellings. On the eastern side of the county, some of the old walls of the houses are of half-timbered construction, with in-fillings of wattle and daub. Examples of even earlier forms of house construction can be viewed at the Avoncroft Museum of Buildings, situated two miles outside Bromsgrove. This is an open-air museum which has reconstructed a variety of buildings, ranging from the Iron Age to modern times. A thatched Iron Age hut, with low stone walls, makes an interesting feature. The conical-shaped thatched roof rises to a sharp pointed pinnacle. The top of the doorway of the hut is built at the level of the eaves of the thatch. The eaves then sweep down to a level nearer the ground, on each side of the doorway.

The main south road from the ancient cathedral city of Worcester passes through the village of Kempsey. Several very old thatched cottages may be seen in the village. Most of the roofs are thatched with long straw and a few display extraordinary tall chimneys. Upton Snodsbury is just off the east road from Worcester, about six miles from the city centre. Some excellent examples of timber-framed properties, with long straw thatched roofs, may be viewed here.

The village of Great Comberton rests at the foot of Bredon Hill, which rises to a height of 961 feet, to the west of the market town of Evesham. The village of Great Comberton contains many thatched cottages and half-timbered farmhouses. Also of interest in the village are the multi-hole dovecots. One of them is believed to be the largest in England. The adjoining village of Little Comberton also has a huge dovecot at the manor house. The village has, in addition, many picturesque thatched timber-framed houses. The nearby village of Bredon again conceals many thatched half-timbered cottages.

Cropthorne lies three miles to the west of Evesham. This village looks exceptionally charming with its long main street of many black-and-white half-timbered cottages with thatched roofs. Cropthorne also has a thatched post office and a seventeenth-century thatched timber-framed farmhouse. The beauty of the spot is further enhanced by its many orchards. The village of Childswickham can be found three miles to the south-east of Evesham. Again thatched timber-framed houses may be seen and also an ancient fourteenth-century cross at the roadside, near the church. The village of Sedgeberow, to the south of Evesham, is also worth a visit, as it contains a good collection of ancient timber-framed thatched cottages.

Offenham is located about two miles to the north of Evesham. This attractive place has many thatched cottages and again some old dovecots. The old-world atmosphere is additionally preserved by a very tall maypole, topped with a cockerel. A few miles westward, at Eastnor near Ledbury, some excellent examples of timber-framed properties, with long straw thatched roofs, may be studied. Bartestree is three miles to the east of the town of Hereford, and long straw thatched cottages are sprinkled throughout the village. A little further westwards, Mansell Gamage approaches Wales on the other side of Hereford. This spot, about eight miles west of Hereford, claims many black-and-white cottages, and most are topped with thatched roofs. Vowchurch, to the south-west of Mansell Gamage, also possesses some thatched cottages. The ancient timber-framed post office has a thatched roof. A bridge spans the River Dore at Vowchurch, near the sixteenth-century Old Vicarage. The village has beautiful surroundings, as it is set in the tranquil Golden Valley.

HERTFORD

The majority of thatched buildings in Hertford are to be found in the western region of the county. A thatched roof o outstanding beauty covers the property belonging to the Ovaltine Dairy Farm, at King's Langley. The thatch has a raised ridge with scallops and points. As a contrast, squares have been skilfully cut under the chimney ridge area of the thatch. The thatched roof completely surrounds the upper small windows of the building, below which are laid raised ornamented thatch aprons. The village of Aldbury rests on the edge of the Chiltern Hills a few miles within Ashbridge Park, to the north-west of King's Langley. Many thatched cottages overlook the village green, which still retains its old whipping-post and stocks. A pond adds further old-world atmosphere to the scene. Other items of interest in the village include a seventeenth-century timbered manor house and some old thatched almshouses which belong to the same period.

Further to the north and about six miles north-east of the town of Stevenage, may be found the village of Cottered. This possesses an extraordinary lengthy row of long straw thatched houses. The walls of the houses are plastered and weather-boarded at their bases. These ground-level weather-boards help prevent the rain splashes, from the eaves of the thatch, spoiling the plasterwork. Dormer windows peep through the thatched roof, which, at the end of the terrace, sweeps down to ground floor ceiling level. The village of Rushden is two miles north. This remote village conceals a group of thatched cottages of exceptional quality and charm.

Charles Lamb once owned a small thatched cottage near the village of Westmill, about eight miles to the east of Stevenage. The cottage, called Button Snap, has been well preserved because of its past literary connection. The thatched roof reaches down to a low level towards the ground; the small windows of the cottage have a diamond pattern. The village also claims a thatched museum which overlooks the inn, exhibiting items associated with the past life of the village. An ancient thatched barn can also be found nearby. The village of Great Wymondley lies three miles to the north-west of Stevenage. There are several clusters of thatched cottages, in varying states of repair, in and around the village. Great Wymondley has many historical royal links, and Henry VIII was once entertained in the village by Cardinal Wolsey.

In the extreme north of the county, the village of Therfield is located three miles to the south-west of Royston. Therfield has many old thatched timbered cottages and also the carefully restored and thatched timber-framed Tuthill Manor. Parts of the latter building are crooked, which gives it a quaint appearance. In addition to the main thatched roof, a small thatched canopy covers a bow window. The village of Barkway stands four miles to the south-east of Royston. Many seventeenth-century thatched cottages line the main street. A thatched antiques shop may also be visited in the village. The hamlet of Nuthampstead Bury nestles close by, containing several picturesque thatched houses. They are thatched with long straw, and the walls are mainly timbered. The small town of Ashwell borders the county of Cambridge, about six miles to the west of Royston. The spring pool at Ashwell feeds the River Rhee. There are many timber-framed buildings in the village. A few of the old houses have overhanging storeys, and some are roofed with thatch. An ancient museum in the town exhibits many agricultural implements connected with the past history of the village; among these is included an array of straw-plaiting tools.

BUCKINGHAM

Chequers, the official country house of the Prime Minister, lies about three miles outside the town of Princes Risborough and close to the villages of Monks Risborough and Ellesborough. These are all located to the south of Aylesbury, and all possess thatched dwellings. Many thatched and timbered cottages stand in the centre of Princes Risborough. Some of the cottages in the town date to the sixteenth century, and there is also an eighteenth-century manor house. The hamlet of Horsenden adjoins Princes Risborough. This contains some lovely long straw thatched houses, complete with ancient dovecots in their gardens. At the nearby village of Monks Risborough, there are many fine examples of thatched and timbered buildings. (It is thought that this village once belonged to the monks of Christ Church Canterbury, before the dissolution of the monasteries.) The neighbouring villag of Ellesborough has thatched dwellings and also offers drinks under a thatched roof, at the ancient Rose and Crown Inn.

To the south-west of Aylesbury, at Long Crendon, can be seen many other examples of thatched cottages constructed in the sixteenth and seventeenth centuries. It also contains a fifteenth-century timber-framed Court House, which was originally given to Catherine of Aragon by King Henry VIII. The property now belongs to the National Trust. Two miles to the north of the village, the road leads to Chilton. Many old cottages warrant study here, including some with walls constructed of wattle and daub. There are also some timbered cottages, and many of the roofs are thatched. In addition, the village offers beautiful scenic views, as it is sited on high ground. Cuddington, about three miles to the east, is slightly unusual in that it boasts two village greens, one of which claims the old village pump. There are many thatched cottages with whitewashed walls in the vicinity.

In the north of the county, the village of Winslow is located just off the main road linking Aylesbury with the town of Buckingham. Winslow is an attractive village with many long straw thatched cottages. Several of these are fitted with large overhanging gables. In addition, the village offers other old interesting buildings. Winslow Hall exhibits many beautiful baroque wall-paintings and tapestries. There is also The Bell Inn which claims that the notorious highwayman Dick Turpin was once a frequent visitor. The nearby village of Hoggeston is well endowed with thatched whitewashed cottages.

A short distance to the north, approaching the town of Bletchley, can be found Newton Longville. This pleasant place features an Elizabethan manor house and many timbered cottages with thatched roofs. Further north, near the county border with Northampton, is the village of Hanslope. This displays many thatched cottages near its very tall towered church, which forms a well-known landmark for many miles over the surrounding countryside.

BEDFORD

There is a great number of thatched properties in the county of Bedford. At Ickwell, a most beautiful village to the south east of Bedford town, there is a collection of thatched houses completely surrounding the very large village green. (The green measures approximately half a mile across.) The thatched houses are unusual in that they are low but still fitted with dormer-type windows. A large maypole stands in the centre of the green, together with an old smith's shop. The village of Old Warden is situated just two miles away. In addition to the many pretty thatched cottages in this spot, a magnificent thatched roof covers the lodge leading to Old Warden Park. The lodge roof is thatched with long straw, and the ridge is richly adorned with scallops and points. The roof is supported on a series of vertical rustic pillars. The lodge building is of a peculiar shape, with the main roof section curved but with straight hip sections at the end overhanging the pillar supports. Another worthwhile place to visit is the Shuttleworth Collection, which can be found on the road linking Ickwell Green with Old Warden. The collection consists of old aircraft and vintage cars, gathered together at an aerodrome which is open to the public.

Clifton is four miles to the south-east of Old Warden. This village displays many good examples of thatched buildings. In the more extreme south-east of the county, the motorway separates the village of Caddington from the town of Luton. However, despite the nearness of the motorway, some thatched houses still surround the village green and the old village pump. The house walls consist mainly of stone and plasterwork, with the thatched roofs descending low and neatly around the upper windows. There are also adjacent houses with tiled roofs.

A rather unusual building is used as the Congregational Chapel at Roxton, seven miles to the north-east of Bedford. The Chapel possesses not only a thatched roof but also a tree-trunk verandah. The building was a barn before it was converted into the chapel, at the beginning of the nineteenth century. Later, two additional wings were built. The roof consists of combed wheat reed, and the decorative ridge has exceptionally lovely scallops and points. A conical-shaped roof section abuts the gabled thatched area and signifies the position of the extension. The village of Roxton rests near the River Ouse; many delightful thatched houses and barns may be seen by the banks of the river.

Melchbourne is about ten miles to the north of Bedford. The village has a whole street of thatched cottages which were erected in the eighteenth century. This spot is also of note as it once formed a centre for the Knights Hospitallers. The main north-west road from Bedford passes through the village of Clapham, which nearly adjoins the suburbs of the town. On the outskirts of Clapham, a quaint thatched cottage overlooks a ford. The St Thomas of Canterbury Church at Clapton has a rather rare large observation tower, constructed during Anglo-Saxon times. A short distance to the north-west, just outside the village of Stevington, a post windmill with working sails may be viewed. A thatched cottage guards the approach to the mill. About another three miles to the north-west, a thatched inn may be visited at the riverside village of Odell. This spot also has many stone-walled thatched cottages. The neighbouring pretty village of Sharnbrook also reveals many thatched cottages. An attractive terrace of dwellings with ashlar walls and thatched roofs stands just off the centre of the village. This long stretch of thatch has an apparently very high pitch. The roof sweeps around and below the level of the dormer windows, which are positioned extremely low, near the floor level of the upper rooms of the cottages. The village of Harrold is also close to Odell. This has many lovely thatched cottages located near the octagonal-shaped Market House by the tiny village green.

A visit to the village of Stagsden, five miles west of Bedford, is always worthwhile, to enjoy the Stagsden Bird Gardens. Well over a thousand birds, many of them extremely rare, are kept in this Bird Zoo and breeding establishment; it is opened to the public throughout the year. The village of Stagsden, near the bird zoo, is only tiny, but it contains several pretty thatched cottages. The main south road from Bedford soon leads to the village of Elstow which has many thatched houses and cottages. John Bunyan was born in the village, and a miscellany of items relating to his life and work can be viewed in Moot Hall. This medieval timbered building has an overhanging upper storey. The town of Ampthill lies a little further south. This preserves many old buildings and past historical connections. Henry VIII and Catherine of Aragon were once frequent visitors to Ampthill. Thatched cottages may be found in the town, and an attractive group exists in Woburn Street. A neat row of thatched semi-detached cottages stands on the Russell estate, built at Ampthill during the earlier part of the nineteenth century. The cottages look nearly identical, and the thatch sweeps over the windows in the end walls in the form of a Sussex hip.

The village of Flitwick rests by the river about two miles south of Ampthill. Several picturesque thatched cottages line the village green. About five miles to the west of Flitwick can be found Woburn Abbey and the Wild Animal Kingdom. Both are open to the public. Woburn Abbey constitutes the stately residence of the Duke of Bedford, and it houses a magnificent collection of paintings and furniture; the Wild Animal Kingdom, amid the surrounding parkland, exhibits many animals in the game reserve. The village of Woburn contains several attractive thatched cottages, some of which are constructed with contrasting red and white bricks. Other old buildings may be discovered in the village, including several ancient inns. A magnificent thatched barn may be seen at Milton Bryan, about a mile away to the south. The barn can be found at the manor house. In the south-west corner of the county of Bedford, the town of Leighton Buzzard borders the River Ouse. This town possesses many old buildings, and among these are thatched cottages constructed of brick and timber.

NORTHAMPTON

Thatched roofs cover many buildings in the county of Northampton, despite the competition of the locally manufactured Collyweston tiles. These roofing tiles are still made in the traditional way in the village of Collyweston, which is situated to the north-east of Kettering.

A village of historic interest in this north-eastern corner of the county is Fotheringhay, where Mary, Queen of Scots, was beheaded in 1587. A row of thatched cottages can be seen very close to the site of the castle where the execution took place. Many of the older cottages in the village have exceptionally small doors. The River Nene flows by Fotheringhay, and further along the river towards Kettering, beside a mill and ford, is Wadenhoe. This village contains a mixture of thatched and tiled roofs. Beautiful countryside surrounds this spot, with several farms and barns in close proximity.

A few miles to the west, the main north road from Kettering runs through the village of Geddington. Thatched roofs protect most of the old cottages in the village, and the post office also boasts a thatched roof. Other items of note in the village are a medieval bridge spanning the River Ise and an ancient stone cross in the square. This was erected on the instructions of Edward I to denote one of the places where his wife's coffin rested during 1290, on its journey to Westminster Abbey. Just before reaching Geddington, the road from Kettering passes through Weekley. This attractive spot also has many thatched cottages.

Rockingham overlooks the River Welland, about seven miles directly north of Kettering. Several old stone cottages with thatched roofs line the long main street of this hillside village. It is possible to obtain magnificent views of five counties by climbing to the top of the hill, when visiting Rockingham Castle. This was originally built by William the Conqueror and was used by many early kings of England. Drastic rebuilding and alterations have been made since that time, and the present building embraces components from nearly nine hundred years of English architectural styles.

The village of Gretton can be found approximately five miles to the north-east. This resembles Rockingham in a couple of aspects. Gretton stands on a hilltop from which outstanding views may be obtained. It also contains several old thatched cottages. Two miles outside Gretton stand the house and gardens of Kirby Hall, which are open to the public. The house was built in 1570, but many alterations were later inspired by the architect Inigo Jones. The main west road from Kettering leads, after two or three miles, to the town of Rothwell. An intriguing thatched building stands on the outskirts of the town. A date on the door suggests it may have been constructed in 1660. It is reached by a series of sharply-rising stone steps. A large thatched roof shelters the building.

Another intriguing but smaller thatched property exists at the village of Flore, to which the main west road leads from Northampton. It is believed that ancestors of John Adams once lived in this thatched building. (John Adams became President of the United States of America immediately after George Washington.) As might be expected, the thatched house is now called Adams' Cottage. It is an interesting observation that the ancestors of George Washington also lived nearby in the county of Northampton, a few miles to the south at the village of Sulgrave. The small Elizabethan manor house, called Sulgrave Manor, attracts many American visitors, as it is open to the public. The village of Moreton Pinkney, two miles to the north, is also worth a visit to see its many seventeenth- and eighteenth-century stone-built houses topped with thatched roofs.

In the extreme south of the county, the lovely village of Aynho sits on a hilltop, with a terrace of thatched stone cottages lining the village square. The occasional thatched farmhouse, with a dovecot incorporated in an end gable wall, may also be discovered. This arrangement is slightly unusual, as most dovecots were built to be free-standing and separate from the house. At Aynho Park, a magnificent seventeenth-century mansion is open to the public.

Another village on a hillside is Badby, which is located to the west of Flore. Badby contains several picturesque thatched cottages and is set amid woods, parkland and lakes. Charles I used to hunt deer in the nearby Fawsley Park. Another park with past hunting connections can be found near the village of Boughton, bordering the northern outskirts of the town of Northampton. Boughton Park contains falconry towers from which hawks were once trained. The village of Boughton also has some pleasant thatched cottages to enhance the scene.

(10)

Thatched Buildings in the North Midlands and the Northern Counties –

Salop, Stafford, Derby, Nottingham, Leicester, Lincoln, Yorkshire, Northumberland, Cheshire, the Isle of Man

SALOP

Salop, or Shropshire, has a high density of well-preserved half-timbered buildings. Many of these are black and white in colour. Several vaunt thatched roofs.

The pretty village of Bucknell, in the extreme south-west corner of the county, borders both Wales and Hereford and Worcester. Bucknell contains a mixture of dwellings, some of stone and half-timbered construction and several with thatched roofs. The north road from Bucknell leads, after two miles, to the hamlet of Bedstone. This also has one or two thatched cottages. In addition, it has a fifteenth-century timber-framed manor house.

In the opposite north-east region of the county, the village of Tibberton lies about five miles to the west of Newport. There are several Victorian-built cottages in the village but also several older dwellings with thatched roofs.

Two miles south of Church Stretton can be found Little Stretton. A black and white church with a thatched roof stands in the village. The church is relatively modern, as it was built at the beginning of the twentieth century. It is therefore unusual for thatch to have been used as the roof material. The hamlet of Minton neighbours Little Stretton.

This is also worth a visit to gain an overall impression of how a small Anglo-Saxon settlement may once have looked. Cottages, farmhouses and manor house cluster in a rough circle around a small green and beside an Anglo-Saxon mound.

The northern road from Church Stretton reaches, after about three miles, the village of Leebotwood. This attractive place yields several pretty black-and-white timbered houses and also a thatched inn. An outside beam, dated 1650, gives a clue to the probable age of the pleasant thatched hostelry. The inn also has some fine Jacobean panelling. The thatched roof is of the winged type and is finished with a neat raised ridge. This is ornamented with a cross rod pattern, and the points cut along the lower edge of the ridge are decorated. The enjoyment of a visit to any of the villages in the neighbourhood of Church Stretton will be further increased by ascending the ridge known as Long Mynd. This runs in an approximately parallel line along the west side of the villages and commands fantastic views over the surrounding English countryside and the Welsh counties. Long Mynd stretches, north to south, for about ten miles and is approximately 1,700 feet high.

STAFFORD

In the east of the county of Stafford, at the village of Alrewas, the wide main street has many thatched houses constructed in Tudor times. Most of the dwellings in the village are black and white in colour. The locality was once very famous for its basket-makers, who used the osiers that grew along the banks of the River Trent. In addition to the river, the Trent and Mersey Canal flows through the village. Walton, which lies near the county town of Stafford, has a post office in the form of a fine thatched cottage. Walton also has an unusual style church with a centre spire and tower. The motorway carves a path along the western suburbs of Stafford. About a mile to the west of this section of the motorway is the small village of Seighford, which has retained many of its picturesque thatched cottages. About eight miles north of Stafford, the motorway also passes the village of Swynnerton. There are several thatched cottages here, together with an old smithy and an inn. A chestnut tree in the village is reputed to be the one which inspired Longfellow to write his poem 'Under the Spreading Chestnut Tree'.

A thatched farm-cottage at Shallowford, five miles to the north-east of Stafford, has been burned down twice this century. After each of these disasters the cottage was rebuilt, refurnished and restored as carefully as possible to recapture its former self. This cottage, known as Izaak Walton's, is open to the public as a memorial museum. It is much visited by anglers who still revere Izaak Walton as the world's most famous angler, although he was born as long ago as 1593. He wrote *The Compleat Angler*, an acknowledged masterpiece, in 1653. Izaak Walton bequeathed the thatched cottage, then part of his farm, to the town of Stafford so that the rent money yielded from it could be put to charitable purposes. Those stipulated included the cost of apprenticing two poor boys, the provision of a marriage portion for a servant girl and finally the purchase of coal for those in need. The cottage is relatively small, with a black and white exterior. It is of timbered construction, and the hipped thatched roof material is long straw.

Izaak Walton was baptized in the Church of St Mary in the county town of Stafford where he was born. This church has a fairly uncommon octagonal tower. The town of Stafford is sprinkled with many ancient buildings. Among these is a picturesque tiny timbered building with a thatched roof which adjoins the seventeenth-century Noel Almshouses in Mill Street. The village of Ellastone settles between Stoke-on-Trent and Derby, to the north-east of Stafford. This spot is much visited for its literary connections with 'Adam Bede', the creation of George Eliot. (Ellastone was the Hayslope of Adam Bede.) Some of the thatched houses identified in the book may still be found in the locality.

DERBY

In the past, flax was often employed as a thatch material in the county of Derby. The walls of the dwellings in the northern parts of the county were mainly constructed of stone, but many in the south were later built with bricks. At the present time, there is a relatively small number of thatched properties remaining in the county.

Two miles north of the centre of Chesterfield is the suburb of Old Whittingham. A small stone-constructed house with a thatched roof, called Revolution House, can be found here. It was once an inn and was then called The Cock and Pynot. In 1688, William Cavendish, the fourth Earl of Devonshire, and his associates used the inn as a meeting-place to plot the downfall of James II and put William of Orange on the throne. The thatched house is now a museum containing a fine collection of seventeenth-century furniture. Osmaston is one of the most pretty villages to be found in the county. It is situated two miles south-east of Ashbourne, near the border with Stafford. The many lovely thatched cottages at Osmaston are set in the delightful background of a park, with a pleasant lake.

NOTTINGHAM

Barton in Fabis sits on the southern outskirts of the city of Nottingham. However, this spot keeps its individual charm, and thatched cottages, together with thatched farms, may still be discovered in the village. The nearby suburb of Clifton is located even closer to the city centre, but again much of its original rural charm has been retained. Thatched cottages line the village green which displays an eighteenth-century gabled dovecot.

LEICESTER

The county of Leicester still has a fair sprinkling of thatched roofs, due to its essentially agricultural history. Many of the houses in the past were timber framed, and the walls were in-filled with stones or clay.

About ten miles south of the city of Leicester, the motorway runs by the side of the rather large village of Lutterworth. Here there are some good examples of timber-framed cottages with thatched roofs. One dates back to the seventeenth century. The village is of note because John Wycliffe, the religious reformer of the fourteenth century, lived there when translating the Bible from Latin into English.

Newtown Linford lies about five miles to the north-west of the city of Leicester, and, due to its nearness to the city, there are many new houses in the locality. However, the village remains one of the most lovely to be found in the county. Several half-timbered cottages with thatched roofs still line the sides of the road. Also, several examples of old timber-framed cruck houses still survive. These picturesque houses are furnished with thatched roofs topped with ornamental ridges. An unusually large number of very small points, similar to the teeth of a saw, have been cut along the bottom ridge edges. Bradgate Park borders the village and has a wooded and rocky heathland terrain. The park is now a nature reserve but still preserves the ruins of Bradgate Hall, where Lady Jane Grey lived during the sixteenth century. On the day of her execution, after being Queen of England for only a few days, the oak trees near Bradgate Hall were severely lopped. Some of these old trees still survive.

Further to the north-west, near the border with the county of Nottingham, rests the village of Hathern. This also has several old thatched cottages. On an historical note, the inventor of the Heathcoat lace-making machine once lived in a thatched cottage here. At the time of its introduction, the machine revolutionized the art of lace making.

About three miles from the northern suburbs of Leicester, lies the village of Queniborough. This has a Norman church with an exceptionally tall steeple. It also has many exquisite old cottages with thatched roofs. The majority are built with the local ironstone. There are several thatched villages in the vicinity of the town of Oakham, formerly in the tiny county of Rutland. Again many of these were built with locally available materials – in this particular region, ironstone or limestone. At the beautiful village of Ayston, to the south of Oakham, there are many cottages constructed with the familiar light brown ironstone and topped with thatched roofs. At Exton, to the north-east, there is a series of limestone-constructed cottages in the village, many of which possess thatched roofs. This village is set amid delightful scenery and by the broad acres of Exton Park. The Church of St Peter and St Paul stands in the park and rewards a visit. It guards one of the finest collections of monumental sculptures to be found in England. Most date from the sixteenth to the eighteenth centuries. The nearby village of Cottesmore contains several delightful stone houses, protected with long straw thatched roofs. The local quarries again yielded the stone from which they were built. Cottesmore is famous as a leading centre for fox-hunting in England. Packs of foxhounds have been continuously retained at Cottesmore since 1788.

At Edith Weston, named after Queen Edith, the wife of Edward the Confessor, are many thatched cottages built with the local stone, and they are all grouped around the village green. The village once belonged to Queen Edith and rests five miles to the south-east of Oakham. At Lyddington, further south of Oakham, the buildings in the village are arranged entirely along a single street. Many of the cottages carry stones marked with dates in their walls, and these reveal that some were built when Elizabeth I was on the throne. Again, most were constructed with ironstone, although a few were built with limestone. A long straw thatched roof still shelters a seventeenth-century farmhouse at Lyddington.

The little town of Market Bosworth is situated in the far west of the county. This old town has several very attractive thatched cottages. Market Bosworth is also of note because of its historical associations. Its name is linked with the Battle of Bosworth in 1485 which terminated the Wars of the Roses. The battle took place a few miles to the south of the town. A place of historical connection with the seventeenth-century Civil War, is Lubenham. Charles I stayed in the village on several occasions during the war. The little village still preserves a few thatched dwellings and is found just outside Market Harborough, in the south of the county.

LINCOLN

Somersby, six miles north-east of Horncastle, was the birthplace of Alfred, Lord Tennyson. The poet was born in a cottage here which was later extended by his father for use as a rectory. Another building of interest, but just outside the tiny village, is an excellent example of a single-storey thatched dwelling. Although constructed in the seventeenth century, as a single-parlour thatched farmhouse, it represents a building which was typical of the long houses built by the Saxons, who shared their living quarters with their livestock. The village of Dalderby is located a couple of miles along the south road from Horncastle. A lovely farmhouse with a thatched roof stands in the village, but it is more conventional and larger than the rather unique one situated near Somersby. A further few thatched roofs can be found sprinkled through the village of Thimbleby, which is two miles to the west of Horncastle. The church here is a little unusual as it is rather large for the size of the village.

The small town of Alford, in the east of the county, has a brick-walled manor house, built in the seventeenth century, with a thatched roof. Another well-known thatched building is The White Horse Hotel. This was originally a sixteenth-century posting-house. Also of interest is the old windmill which still stands in the town. The north-east road from Alford leads, after a couple of miles, to the village of Markby. This possesses a very old thatched church which has been well restored.

In the northern region of the county, the village of West Rasen is about three miles to the west of the town of Market Rasen. A thatched post office and general store may be visited in the village. West Rasen also claims a packhorse bridge which was built during the fourteenth century. South Willingham is situated to the south-east of Market Rasen, and a few lovely thatched cottages add considerable charm to the village.

The town of Spalding rests by the River Welland in the south of the county. Delightful fields of bulbs and flowers, reminiscent of Holland, surround the town and attract many visitors in springtime to admire the multitude of blooms. The town of Spalding conceals many old and fascinating buildings. The White Horse Inn shows an attractive thatched roof. The village of Fleet is located to the east of Spalding, in the direction of The Wash. The rather unusual church at Fleet has a separate tower and spire. A thatched lych-gate guards the path to the church, which dates back to the fourteenth century.

YORKSHIRE

Very few thatched roofs remain in Yorkshire. This is mainly due to the great popularity of pantiles, after they were introduced from Europe at the beginning of the eighteenth century. Most of the surviving thatched roofs are to be found in North Yorkshire.

Runswick is on the North Yorkshire coast, near Whitby, the town of Captain Cook. Despite its apparently exposed position, Runswick gains good shelter from the northerly winds by the tall cliffs at the extremity of the bay. This sheltered situation has made Runswick Bay a popular holiday resort. It has also made it a practical proposition to roof some of the houses with thatch. A few still survive, in admixture with the red pantiles roofs.

Further thatched dwellings exist in some of the villages surrounding the pretty market town of Helmsley, bordering the Yorkshire Moors. The Farndale area to the north-east contains some very old cottages with thatched roofs. The Farndale Valley forms a well-known beauty spot and nature reserve. Thousands of wild daffodils grow in the confines of this protected moorland locality. The nearby village of Hutton-le-Hole still retains the occasional ancient thatched cottage, but most of the stone houses constructed in the seventeenth century are red roofed. This delightful moorland spot has a stream running through the centre of it, and small white bridges span the sparkling water. The Ryedale Folk Museum, near the local inn, exhibits a miscellany of items connected with the rural life of the region. The village of

Pockley, much closer to Helmsley, also displays a sprinkling of thatched roofs. Harome, to the south-east of Helmsley, likewise still has a few old thatched cottages.

The racecourse, just outside the medieval city of York, boasts a thatched building, standing near the winning post. It takes the form of a small construction which is raised clear of the ground by vertical supports to give it good observation. The hipped roof is thatched with long straw and finished with a neat ornamented raised ridge. A single ligger runs along and secures the bottom eaves level of the thatch.

Warter, to the east of York in Humberside, is a quaint village with several thatched cottages. Also of special interest in the near vicinity are Burnby Gardens at Pocklington. These are peaceful gardens with ponds growing some of the finest specimens of water-lilies to be seen in Europe. They are open to the public through the summer months. There is also a museum in the grounds of the Hall.

NORTHUMBERLAND

In the extreme north-east of England, at Etal near the Scottish border, an unusual street of thatched cottages can be seen. A particular feature is that at one end of the street an eighteenth-century manor house stands, while at the other there remain the ruins of a castle destroyed in 1496 by James IV of Scotland. The village also contains a lovely inn with a thatched roof. Most of the buildings are constructed of stone with whitewashed walls. It is very rare to find thatched roofs in such a village sited so far to the north in England. However, Etal has retained its old thatched roofs, in a mixture with others of slates and pantiles. It is perhaps noteworthy to relate that thatch was the traditional roofing material used in the locality of Etal during the eighteenth century. Sandstone slabs were generally utilized in most other areas of Northumberland.

CHESHIRE

A wood of Scots pines, belonging to the National Trust, overlooks the village of Burton. This charming place is situated seven miles to the north-west of the ancient city of Chester. Many white-walled half-timbered cottages with thatched roofs stand in the village. Charles Kingsley wrote about this locality in his poem 'Sands of Dee'. The neighbouring hamlet of Ness is also historically well known, as it was the birthplace of Nelson's Lady Hamilton. The hamlet of Raby is a further three miles to the north. This offers hospitality at an early seventeenth-century thatched inn, The Wheatsheaf, a half-timbered building with dormer windows; its thatched roof has been well maintained and is ornamented. A local beauty spot known as Raby Mere hides close by, screened by woods. A thatched timbered millhouse, with a raised ornamented ridge, overlooks the quiet stretch of water at this delightful area, which is popular at weekends for boating and picnics. The Norfolk reed thatched millhouse was built during the same period as the inn at Raby. Many old thatched cottages still remain in the Bidston region of Birkenhead, to the north of Raby. Also of note is Bidston Hill, which has been conserved as an open heathland space for the public to enjoy. An old windmill makes a prominent and familiar landmark in this area.

Several half-timbered thatched cottages, mainly black and white in colour, can be found in the tiny village of Peckforton, approximately ten miles to the south-east of Chester. A strangely ornamented stone beehive may be spotted in one of the gardens of the cottages. The beehive is carved in the shape of an elephant, with a 'castle' on its back. Another intriguing place, to the north of the village, is a nineteenth-century castle set on a hill. It was designed by Salvin. Further delightful thatched cottages may be seen at Norbury, to the south of Peckforton.

South of Knutsford, at the small village of Lower Peover, there are many cottages constructed of timber and roofed with thatch. An old watermill is preserved close by, together with an unusual church with large overhanging eaves and black-

and-white gables. An ancient inn stands by the church. About eight miles to the north, the town of Hale, near Altrincham, now forms a part of Greater Manchester. Despite the encroachment of modern town buildings, several old properties still survive. One such building is Hope Cottage, located in Bank Hall Lane. This thatched timber-framed cottage was built in the sixteenth century, and during its long history was once used for a short time as a mission. It is furnished with a lovely thatched roof, with a raised ridge and neat points cut along the bottom edge. The ridge and points are immaculately ornamented with diamond-shaped wood patterns.

THE ISLE OF MAN

The Isle of Man is located about thirty miles off the west coast of northern England and is much frequented by English holiday-makers. The Isle of Man warrants a brief mention here as broom was once utilized as a thatch material to protect the Manx cottages, constructed with stone, mud and clay walls. Later straw was used as the thatch material, and the roof was then bound with straw ropes, fixed to the walls, to hold the thatch firmly down. Sometimes, heavy stones were secured to the ends of the ropes to weight them, rather than employing the alternative wall-attachment method. A few old Manx dwellings still remain on the island.

Cregneish sits on the southern tip of the Isle of Man and is a delightful village with many old thatched cottages. Due to its lofty position, the spot also offers stupendous views. The Manx Village Folk Museum at Cregneish consists of a number of thatched cottages and workshops. They depict many facets of the old Manx way of life. The items exhibited include many associated with agriculture, weaving and furniture manufacture. The thatched roofs of the museum buildings have been preserved and renovated in the style identified with the Isle of Man.

(11)
Costs and Modern Trends in Thatch

Many of the old thatched cottages described in the last few chapters cost very little when they were first built. Labour and materials were both cheap. This sometimes causes visitors from Third World countries to view our delightful cob-walled, thatched cottages with scepticism. This is because of the original use in their construction of cheap local building materials obtained from the earth and they associate them, incorrectly, with the simple mud huts with grass roofs that exist in their own villages. They forget that most of our thatched cottages were constructed robustly enough to have survived several centuries. Also, like all good antiques, they have considerably matured in value with the passage of time. Many of these once humble farm-labourers' cottages are now 'highly desirable residences'.

Estate agents report that properties with thatched roofs, now command a considerable price premium (10 per cent or more) over houses of the same period in similar condition with tiled or slated roofs. This apparent current trend and preference by purchasers for thatched homes outweighs the fact that considerable extra, although not prohibitive, expense will be encountered in maintaining and insuring them. However, it is only prudent to buy a thatched property if the new owner can afford this extra outlay. There are few sights more forlorn than unkempt thatch that has become bedraggled by neglect.

The present demand for thatched cottages has encouraged some building firms to construct more new thatched homes. Several have been built in expensive residential areas in the 'stockbroker' belts as exclusive family houses. Although the thatched roofs on new houses always look attractive, the buildings themselves can never achieve the quaint charm of the old cob-walled cottages with their uneven walls, floors and patinated crooked beams.

On the question of the cost of completely re-thatching an old property, it is always difficult to be precise. For example, a small cottage nowadays may cost between £2,000 and £5,000, a large cottage may cost in the region of £10,000 to £15,000, whilst a very large area of new thatch on a barn may cost between £20,000 and £25,000. The total area to be thatched mainly determines the price but the type of thatching material to be used also affects the cost. For example, if the price for replacing a long straw roof is £5,000, then the cost using combed wheat reed would be about £6,000, whilst in Norfolk reed the price would be approximately £8,000. The number of dormer windows and roof valleys, and the distance the thatcher may have to travel to reach the house also affect the price. The construction of an elaborate ridge also adds to the expense, as mentioned in Chapter 3.

The approximate present-day price of fixing a Norfolk reed roof would probably be about £55 per square metre, whilst combed wheat reed thatch, although cheaper, would still be just over £40 per square metre. A long straw roof would be slightly less expensive at about £35 per square metre. A decade or two ago, Norfolk reed cost nearly double the price of combed wheat reed but due to the increasing shortage of long stemmed wheats the price differential has now decreased.

To obtain a rough estimate of the area of a thatched roof two main measurements are required. The first is the vertical distance of the thatch from wall to wall when measured over the top of the roof. The value obtained is therefore the sum of the two slopes from eaves to apex, plus the actual thicknesses of the eaves. The second measurement needed is the horizontal length of the thatch at the eaves level and this should also include the thicknesses of the barges at the ends. The two measurements multiplied together give a good approximate estimate of the main roof area. Knowing the number of squares metres, a rough idea of the cost of a re-thatch may be calculated by the house owner.

However, it is unlikely that a full re-thatch will even be suggested as necessary after a roof examination by a Master Thatcher. In the case of combed wheat reed and long straw roofs, it is more usual to give a new top-coat of thatch over the old, after it has been prepared back to a firm base. This considerably reduces the expense of the work, as material costs will be diminished by about 30 per cent and labour costs will also be less. In the case of a water reed roof, it is more likely that the whole of the old thatch material will have to be removed and a new thatch put on the roof. However, this will then have a much longer life compared to the straw roofs re-thatched by the top-coating process.

In fact, some thatchers think that it is difficult to achieve a real 'craft job' on any roof, unless all the old thatch is first removed. Even if this argument is valid, it seems a shame to strip sound underlying thatch material unnecessarily, especially as it may have been in place for centuries without deterioration. One disadvantage of never stripping the thatch completely is that the underlying rafters and battens may eventually become split and not be repaired.

The Masters' Thatchers' Associations, in the various counties, are always available to help the property owner decide the best course of action to prolong a thatched roof's life. It may be that only minor patching will prove necessary. In the case of a straw roof this may entail repairing bird holes and fastening fresh straw with spars into small damaged areas. In the case of a reed roof, the repair work is usually different. The thatcher draws down the reed in the damaged regions and inserts small bunches of new reed of the correct length. The new reed is securely fixed and a leggett dresses the old and new reed interfaces together, so that the repair assumes the same level and homogeneity as the main roof. The passage of time soon harmonizes the different colours of the old and new reed.

The only repairs a householder would be wise to attempt himself are those to bird holes in the eaves or barges. These do not involve the need for a ladder to be placed on the thatch with the possible risk of causing further damage. Even these minor repairs are only advisable if a lengthy wait for, or difficulty in obtaining a thatcher are encountered.

A bundle of long straw is gathered together of an estimated size to make a good tight fit in the bird hole to be repaired. The bundle is then tied together (loosely at first) with a rot-proofed or tarred twine at two points, roughly at the middle of the bundle and a third way down from the 'ears' end. A half-metre length of willow or hazel, about the thickness of a pencil and with one sharpened end, is then pushed down the centre of the straw. Half the length of the hazel stick should be left protruding from the straw bundle at its 'ears' end. Also, the protruding end of the stick must be the sharpened one. The twine around the straw bundle is then fully tightened. The bundle of straw (protruding stick-end first) is then pushed into the bird hole. A flat surface, such as a cricket bat or the back of a spade is used to tap the bundle more firmly into the hole. The ends of the straw bundle are trimmed to make them assume the same level as the surrounding eaves or barge surface.

An annual inspection of a roof will probably cost around £15, whilst a full survey and written report by a Master Thatcher will cost £40 to £100. It is normally recommended that the ridge on the roof should be renewed about every twelve years. Also, in the case of a long straw thatched roof, it will probably be necessary to replace the protecting wire mesh every ten to fifteen years. Much depends upon the local environment. In this respect, thatched porches demand more attention. The small roofs take the full flood of water cascading from the main roof and they therefore endure harsh treatment from the weather.

On the subject of ridges, a plain type in straw or sedge, keeps the cost down and also forms the best choice for a really old building. Prints and pictures of thatched cottages from the sixteenth to the eighteenth centuries show that traditional ridges were never ornate. In the Middle Ages, ridges were normally covered with clay or turves. Some thatchers think that plain ridges are also longer-lasting and therefore more weather-proof in the long term. However, many people delight in having an elaborate fancy ridge and are quite prepared to break from tradition. Most of the richest decorative ridges are to be found in Norfolk and Suffolk.

It is worth recalling that it is not usually prudent to employ an inexperienced semi-skilled thatcher to carry out any work, despite the attraction of the lower cost and the avoidance perhaps of a lengthy wait for a Master Thatcher. Expert quality thatching depends upon the conscience of the thatcher and his very long experience making him a perfectionist. A Master Thatcher is much more interested in achieving a perfect finish than hurrying on to another job, or having a tea-break. Even if the many semi-skilled thatchers now operating produce a nice-looking roof, the final judgement still lies in how many years it will serve as a good weather-proof roof, before faults develop. Unless the thatch material was expertly applied, the life will be limited. As many semi-skilled thatchers have now become itinerant, it may be difficult to get them back to correct faulty workmanship.

Another subsidiary influence of the travelling semi-skilled thatcher may be the eventual decrease in the use of long straw thatch, in areas where traditionally the material has been employed for many centuries. The handling and preparation of long straw entails more time and trouble. It therefore daunts the itinerant worker, who may suggest to an unwary owner a change to combed wheat reed. Another reason for the intrusion of combed wheat reed into established long straw regions, is the fact that combed wheat reed is more robust and easier to transport. The same, of course, applies to water reeds which are even tougher and are often taken into traditional combed wheat reed areas. Local Master Thatchers in these regions often possess CoSIRA certificates in both water reed and combed wheat reed. Whatever the reason, this alteration in the type of thatching materials, especially a change from long straw, gives a cottage a completely different appearance and with an old building, which is 'listed', it should require listed building consent.

As well as the intermittent maintenance cost, the thatched house owner has to meet the annual extra expense of insuring his home. Unlike the past, most major insurance companies will now insure thatched buildings but all charge special premiums, because of the increased fire hazard. These may range from 50p to £3.50 extra, on top of their standard buildings cover rate per £1,000. For several years, this latter rate has been held at an average premium of £1.80 per £1,000 of cover. Due to the damage caused by the hurricane in October 1987, many insurance companies have increased the rate to £2 or £2.15 per £1,000 during 1988. Some insurance companies differentiate between a straw and a reed roof. This is a recent trend and has developed because the latter type of thatch is considered less vulnerable. Statistics have suggested that the more tightly packed reed roof is marginally safer, as regards the rate of spread of fire.

The Country Gentlemen's Association, Icknield Way West, Letchworth, Herts., has for twenty years been a pioneer in designing special comprehensive policies for thatched property owners, underwritten at Lloyd's. The premium offered depends upon the result of their free survey of the property. This verifies that all electrical wiring and chimney constructions are in sound condition and that the thatch has been maintained to a high standard. A company specializing in thatched property cover is the Thatchowners (Insurance Agency) Limited at Victoria House, 38 Hampton Road, Twickenham, Middx. Another useful address to contact regarding insurance is the Thatching Advisory Service, Rose Tree Farm, 29 Nine Mile Elm Ride, Finchampstead, Wokingham, Berks. Again these offer Lloyd's backed cover. The Thatching Advisory Service also give expert advice on all thatched property requirements and can arrange surveys and free quotations from your local Master Thatcher.

With regard to safety and electrical wiring in a thatched house, it is important to keep small pieces of straw in the loft space away from electric junction boxes, in case they have faulty covers. It is also prudent to use bulkheads to protect electric light bulbs from contact with falling straw debris. As a safety precaution, the lights should always be switched off when the loft is vacated. The fitting of heavy gauge aluminium barrier foil under the thatch prevents debris from falling and also acts as a fire resistant barrier in the event of a fire. It functions further as a thermal reflective insulator.

In the past, the potential vulnerability of thatch to fire has always been recognized, especially in the towns. In the countryside, on an historical note, it is interesting that in the heyday of the coaching era thatch was rarely used on toll houses. A few thatched ones have survived to the present day but toll houses were notoriously unpopular and firing by an angry mob was not uncommon.

Most old thatched buildings have draught-free roof spaces, as the eaves are always closed off. This is good as regards potential fire risk. The lack of ventilation also makes the loft warmer and prevents the condensation problem which would occur in a cold roof space. In the unfortunate event of a fire. all doors should be shut, if it is possible and safe to do so, before immediately leaving the building. No attempt should be made to remove burning thatch from the roof before the arrival of the Fire Brigade. Such action would allow air to penetrate below the thatch and rapidly increase the rate of spread of the fire.

Any wire netting on the thatch will be removed as quickly as possible by the firemen, by pulling the sections apart at the apex and then down the seams. The individual lengths of netting would have been laid not to overlap at their edges. They would be joined together every nine inches, by special pull-apart metal hooks.

The reduction in fire risk by the application of fire-retardant salt solutions, injected into the thatch, has the disadvantage that weathering gradually decreases their effectiveness. The area near the ridge is particularly prone to this and most of the salts are probably leached out within a period of ten to fifteen years. It is also thought that fire-retardant preparations may reduce the life of the thatch. To improve the efficiency of fire-retardant salts, it is best if they are applied to the thatch material before roof construction commences. In the case of straw thatch, the thatcher always wets the straw before fixing it to the roof and he can therefore conveniently substitute a water solution of the fire-retardant salts for this purpose.

In general, the life of a thatch depends mainly on the skill of the thatcher, the type and quality of the material used, whether the roof is in a shady or sunny position and also on the geographic location of the house. It seems a fact that the life expectancy drops the further to the west of the country the property is situated. It has been claimed that a Norfolk reed roof lasts for sixty years in East Anglia but survives for only thirty years in the wetter West Country or Lake District. This may be partly due to the difference in pitches of the roofs found in the east and west of the country. However, it is likely that the climatic conditions play a more important role. The higher average temperature, humidity and rainfall experienced in the west, together with the generally less polluted air encourage the multiplication of minute fungi and biological growths which start the thatch decomposition process.

Modern chemicals, such as heavy-metallo organic compounds, act as fungicides for such growths when applied to thatch. However, they are difficult to keep on the surface of an existing old roof. The roof design ensures that any solution cascades off it quickly. Another disadvantage is that as the fungicide kills the biological growths, they wither and are blown or fall away from the roof. This process removes some of the toxicant with it and so the effectiveness is gradually reduced. A repeated treatment is therefore needed. Fungicides are best added to the new thatch material before roof construction. Any fungicide or spray should only be used when rain is not expected for several days. The solution must have ample time to dry and deposit its active component on all surfaces of the thatch material. The handling of sprays and chemicals requires caution to avoid skin contact and inhalation of mist.

Chemical sprays can also be considered for the killing of moss and lichens which often develop later on old thatch. Suitable treament lasts for several years. However, the physical removal of moss by a thatcher is probably more beneficial. The process is a skilled job and the precise method depends on the thatch material and its state of deterioration. After ridge repair work, a thatcher will sometimes fit a long strip of copper, commonly hidden underneath a ligger, on each side of the ridge. This method is not always successful and depends on future 'weathering' producing toxic compounds which inhibit further growths on the thatch.

An effective long-lasting bird repellent which could be sprayed on a long straw thatch has not yet been satisfactorily marketed. The repellents available remain active for only short periods. Thatchers still rely on a three-quarters of an inch mesh of wire netting, or sometimes plastic P.V.C. netting, to discourage birds. The use of a smaller mesh size, such as a half or five-eighths of an inch, would be even more efficient in deterring birds, especially when the netting under the eaves is attached snugly to the house by a timber fillet. Nevertheless, this size should not be employed for other reasons. It impedes water flow from the roof, traps more leaves and does not allow the thatch to 'breathe' and dry out quickly. In essence, it encourages rot.

Several strange ways of dealing with the sparrow problem on thatch have been recorded over the years. One amusing Irish country method dates back to 1885, when Irish whiskey could reputedly be purchased at thirty-six shillings for a case of a dozen bottles. It seems the technique was to take several large handfuls of wheat and soak them in Irish whiskey for twenty-four hours. The treated wheat was then placed, at intervals of two or three yards, at stategic points on the thatch. It was reported that after about five to ten minutes the birds became helpless and could easily be collected. It is not clear whether the hapless birds were killed or allowed to recover with hangovers, vowing never to visit the thatch again.

On the general subject of bird damage, the problem appears to have increased in recent times. This is perhaps understandable as more trees are felled and hedgerows replaced, thus reducing the insect population, seeds and other food available to them in the open countryside. Bird scarers placed on the tops of thatched roofs are not a practical proposition to solve the problem. The hostile effigies need to be changed frequently, so that the birds do not become familiar with them and lose their fear.

With regard to the building of new thatched country homes, it is hoped that architects will continue to consider thatch as a practical roofing material well into the next century. It is not only valuable for its excellent insulation and energy conservation but it is also aesthetically pleasing. It would be laudable if this latter advantage could be favourably considered when new housing developments are planned in the countryside. This would then help to preserve a lovely rural heritage, despite the higher building costs involved. In addition, thatched villages and hamlets always attract tourists, especially in the West Country and East Anglia. It would be reassuring to think that some of the new thatched buildings being designed and built today, will one day delight future generations. Fortunately, some old redundant thatched agricultural buildings are now being renovated for residential purposes. For example, many thatched barns have been adapted to create new homes in several parts of the country.

Existing old thatched cottages were built long before modern Building Regulations came into force and they therefore do not comply with them. However, new thatched homes now have to be constructed in accordance with the current regulations. They are mainly concerned with the possible spread of fire due to the thatched roof. This restricts the building of new domestic thatched properties to single detached, or semi-detached houses as long as the total cubic capacity does not exceed 1,500 cubic metres. Also, no part of the main thatched roof must measure less than 12 metres from any point on the property's boundary. If a thatched porch or bay window protrudes from the house then this small area of thatch is usually exempted, so long as its total area does not exceed 3 square metres. It must also be separated from the main roof by at least 1.5 metres of a non-combustible material and no part of the protruberance must be less than 6 metres from the property's boundary. These precautions are designed to prevent the spread of flames to a neighbouring building in the event of a fire.

Modern roofing felt is not recommended for use under combed wheat reed or long straw thatch, as these materials are always fixed to the roof in a damp state. The felt impedes the drying out process of the underlying thatch layers.

Roofing felt may be satisfactorily laid under a water reed thatch, as the reed is always applied to the roof in a dry condition. However, the felt placed on the rafters, below the thatch, must be fixed loosely to allow ventilation at the eaves as required by the Building Regulations. This provision is made with new buildings to prevent condensation in the roof void.

In the case of a new thatched house, this ventilation obviously leads to a colder roof space than with an old traditional thatched building, constructed with completely draught-free eaves to give a closed roof-space. Unfortunately, the eaves ventilation required in a new house reduces to some extent the full potential of the thermal insulation capacity of the thatched roof, achieved by its great depth of many small pockets of air trapped between the inner and outer surfaces of the thatch.

The present-day practice of laying felt beneath roofing materials is not entirely a modern innovation. In the early seventeenth century, it was quite common to use inner linings beneath water reed thatch. The lining frequently consisted of a two-inch layer of reeds, bonded together by plaster. The reeds were employed as a key for the plastering. The lining was also frequently covered with several layers of lime-wash. This latter treatment no doubt served a fire-retardant purpose, as well as offering a decorative advantage.

Later, it became fairly common in East Anglia to place woven mats of reed on the rafters below water reed thatch, especially when the roofs were not fully battened. The woven mats gave a neat attractive appearance to the inner surfaces of roofs with exposed rafters, such as barns. The ceilings of thatched verandahs of lodges and houses were also often given a decorative underlining. This consisted of reeds arranged in elaborate radial patterns, with added designs made in fir bark attached to the underside of the reeds. In the West Country, wattle hurdles sometimes supported the thatch. All these methods bear similarities to the medieval practice of using oak laths to support thatch. The laths were made by splitting the oak, rather than sawing it. The laths were also sometimes crudely plastered.

Today, with new thatched buildings the structural members of the roof must conform to the mandatory requirement of the appropriate Schedule of the Building Regulations. Nevertheless, it is essential that the rafters are always more than 50 m.m. thick, so that they can securely hold thatch hooks. The size of the wooden scantlings selected, appropriate to the span, must be able to bear a thatch load which is calculated to be less than 25 k.g. per square metre. Thatch is fairly light when compared to alternative roofing materials. It therefore needs relatively little sophisticated carpentry for the underlying roof structure. The size of the battens recommended for thatching is usually 38 m.m. x 25 m.m. or 40 m.m. x 20 m.m. They are preferably treated against future rot and wood worm attack. Master Thatchers are always willing to offer advice on builders' timber work, both old and new. They can also help by providing useful suggestions on new proposed extensions to existing thatched houses.

The future availability of good quality straw for thatching must depend upon farmers continuing to grow some wheat especially for the thatcher. This means growing it for its straw production, rather than its grain. It is hoped that the present huge surplus of grain and cereal production in the EEC may encourage more farmers to grow some acres of the older longer stemmed varieties of wheat. The market price of quality tough wheat reed and long straw for thatching may reach such a level that it makes an attractive cash crop for some farmers, despite the extra trouble and labour involved. The labour problem on a small modern farm may prove the biggest obstacle, as a team of about twelve people will be needed for the combing and tying operation. The risks to the farmer are also greater, as the crop would be more easily flattened and damaged by bad weather.

In essence, the work entails growing the special varieties of longer straw wheats which must be autumn sown. This ensures the crop has an extended growing period to reach the height and strength required for thatching. Spring wheat is therefore not suitable. Also, nitrogenous top dressings must not be applied the following spring, in an attempt to increase the yield of the autumn sown wheat. The wheat has to be harvested in the summer with a reaper and binder, before being allowed to ripen fully in stooks. It will then probably have to be placed in a rick, before eventually either being combed out for the reed or threshed in a special drum for use as long straw.

Only then will the reed or long straw be of the excellent robust quality desired by the thatcher, so that he can produce a roof with a long life. The reputation of the thatcher suffers if he uses poor quality long straw or wheat reed and he takes the blame rather than the grower. Master Thatchers' Associations are constantly investigating and discussing with farmers the need to secure better supplies of good quality materials. This aspect is essential to their high standards of craftsmanship, so that their artistic trade may continue to flourish.

If the future demand for thatching straw exceeds available supply then it is likely that water reeds may intrude still further into traditional straw areas. This will dull even more the previous sharp regional distinctions. Norfolk reed alone will not be able to meet the shortfall, so the continuous development of other sources of marsh reeds will be essential. However, to reiterate, it is only the common reed, *Phragmites Communis*, of the various species of aquatic grasses that will be suitable. Unfortunately, labour may be difficult to find for cutting these reeds in the winter and increasing water reed acreages are now being conserved for bird sanctuaries. Reed beds form the natural habitat for such birds as reed and sedge warblers, together with bearded tits. The drainage of marshland for urbanization schemes, farming and tree planting projects have also diminished the areas of fens and marshes.

Despite this, it remains a fact that the reeds still grow prolifically in the wild. As indicated in Chapter 2, they still thrive in several areas of the United Kingdom and Europe on unclaimed marshland of inland water and tidal estuaries. In other regions of the world, the reeds also commonly grow and the dried stems of *Phragmites Communis* reeds have been used for arrows, basketry, pens and the manufacture of musical instruments. However, the reeds need to be regularly farmed to make them suitable for thatching. In general terms, the 'wilder' the reeds the weaker will be their basic stem structures. It is difficult to estimate the potential lives of various unproven marsh reeds when incorporated into a roof but at least they offer a possible future supply for the thatcher.

The semi-cultivation of marsh reeds to improve their quality entails cutting the suitable length reeds and then burning the old growths to destroy debris and weeds. The burning process creates potash to encourage new straight reed shoots to grow. Nowadays, special hand-operated machines, such as power-driven scythes, cut the reeds which are then tied into bundles. Mechanical self-propelled cutters are also available. A machine can cut several hundred bundles of reeds per day. The special machines sever the reeds in such a way as to minimize damage to the stalks and to allow the reeds to be easily gathered and bundled.

Other modern aids to the marsh reed farmer include the use of sledges towed by tracked vehicles. These haul the reeds away from the cutting areas, across the marshy ground, to reach lorries parked on firmer terrain. Heavy lorries, loaded with reed, would sink and become immobile in the boggy ground near the reed beds. If the home supply of water reeds should eventually be seriously depleted, increased supplies may be imported from European and East European sources. The reeds are extremely tough and stand up well to transport. However, they are bulky by nature and the cost of carriage will be high. Also, their quality will possibly never rival the reed grown and harvested in Norfolk, acknowledged to be the finest in the world. The reputation for Norfolk reed was gained by many years of continuous cultivation, so it is only possible for other sources to compete by careful husbandry over a long period of time.

A further necessity to the future of the thatching craft will be for farmers to continue growing hazel coppices on their land. This will ensure an ample supply of the wood from which the thatching spars can be fashioned. It is estimated that each thatcher's basic square, or 100 sq.ft. of thatch roof surface, requires about 300 spars. This illustrates the very large number required by the whole thatching industry.

Index